The Facts On File
DICTIONARY OF BIOTECHNOLOGY AND GENETIC ENGINEERING

The Facts On File

DICTIONARY OF BIOTECHNOLOGY AND GENETIC ENGINEERING

Mark L. Steinberg, Ph.D.
Sharon D. Cosloy, Ph.D.

Edmund H. Immergut, Ph.D.
Series Editor

Facts On File®

AN INFOBASE HOLDINGS COMPANY

Library of Congress Cataloging-in-Publication Data
Steinberg, Mark (Mark L.)
    The Facts on File dictionary of biotechnology and genetic
engineering / Mark L. Steinberg, Sharon Cosloy ; Edmund H. Immergut,
series editor.
        p. cm.
    ISBN 0-8160-1250-4 (acid-free paper)
    1. Biotechnology—Dictionaries.   2. Genetic engineering—
Dictionaries.   I. Cosloy, Sharon.   II. Title.
TP248.16.S84   1993
660'.6'03—dc20   93-25337

To Lisa and Katherine (M.L.S.)
To Ed, Rebecca and Michael (S.D.C.)

CONTENTS

PREFACE

We are in the midst of a dramatic revolution in the field of biology in which, for the first time, science is faced with the possibility that change in the genetic makeup of higher organisms might be directed in the laboratory rather than by the random forces of natural selection. This new era was born out of certain critical discoveries in the mid-1970s that led to the appearance of new fields of molecular genetics variously known as "gene cloning," "genetic engineering," and "biotechnology." The central theme of genetic engineering is the introduction of the genetic material of humans and animals into microorganisms where it is then possible to isolate, study, and ultimately engineer individual genes for a variety of specialized purposes in the laboratory. Genetic engineering has already given rise to technologies that were unthinkable barely two decades ago: the cloning of genes responsible for genetic diseases, the synthesis of unlimited quantities of therapeutic agents, human hormones, and critical blood factors in bacterial "factories," the creation of genetically engineered plants and animals, and the ability to derive genetic fingerprints from microscopic samples of tissue—only a few examples of technologies that have been realized even at this writing.

Recently, much of the research in biotechnology and genetic engineering has moved from the academic world into the industrial. As a consequence, many new and potential applications have come into the hands of private enterprise where, fueled by more substantial funding and motivated by market forces, the development of new products has reached an explosive pace. This has also meant that even as the rapidly increasing rate of progress taxes our ability to keep up with new developments, there is an ever-increasing need to understand the legal and ethical issues that inevitably accompany any new technology. However, in contrast to other new technologies, the products of genetic engineering deal directly with fundamental biological processes and are, by their very nature, certain to have an immediate and profound impact on all areas of human health.

This dictionary attempts to provide the basic vocabulary of modern biotechnology and genetic engineering so that those with even an elementary knowledge of basic biology and biochemistry will be able to follow the flood of fast-breaking developments in biotechnology and genetic engineering that constantly appear in the media. Since this book addresses an audience from diverse backgrounds and covers a broad field, we have attempted to include both basic and technical terminology in a number of areas, including plant and animal biology, to meet the needs of as many readers as possible. We have also tried to make the dictionary self-contained in the sense that, in cases where technical terms appear in definitions, these terms are defined elsewhere in the book. Thus, it is anticipated that the dictionary will be of benefit to a wide-ranging audience, including high school and college students, lawyers, physicians, scientists, teachers, librarians, or others with a particular need to keep abreast of the rapidly developing area of biotechnology and genetic engineering.

A

ABO blood group A system of AN-
TIGENs expressed at the surface of human red blood cells. Human blood types
represented in this group are A, B, AB,
and O, depending on the antigen(s), in
the form of oligosaccharides, present at
the surface of the erythrocyte membranes. The blood serum of type A produces anti-B antibodies, that of type B
produces anti-A antibodies, and AB serum produces both. Type O produces
neither. The ABO blood group is one of
14 blood group systems, consisting of
100 different antigens. This system is
medically important because recipients
of a blood transfusion must receive blood
that is compatible with their own type.
Persons with type AB blood are universal
acceptors, and those with type O are
universal donors. The ABO system can
also be used in paternity suits to possibly
rule out that a particular male is the
father of the child in question.

abscisic acid A plant hormone, lipid
in nature, synthesized in wilting leaves.
It counteracts the effects of most other
plant hormones by inhibiting cell growth
and division, seed germination, and budding. It also induces dormancy.

absorbance A measure of the light
absorbed by a solution or a suspension
of bacterial cells. It is a logarithmic function of the percent transmission of a
wavelength of light through a liquid and
is measured by a spectrophotometer or
a colorimeter. Absorbance values are
used to plot growth of suspensions of
bacteria and to determine the concentration and purity of molecules such as nucleic acids or proteins in solutions. Often
referred to as optical density.

absorption (1) *virology* The entry
of a virus or viral genome into a host
cell after the virus has adsorbed to
the cell surface. (See ADSORPTION.) (2)
photometry When light is neither
reflected nor transmitted, it is said to be
absorbed. Some biological systems can
use light energy because they have pigments that absorb light at specific wavelengths. These pigments are able to
harness light energy to drive biochemical
reactions in vivo. An example is plant
pigments, such as chlorophyll, used to
trap light energy and drive the process
of photosynthesis whereby plants manu-
facture nutrients.

absorption spectroscopy Using a
spectrophotometer to determine the
ability of solutes to absorb light through
a range of specified wavelengths. Every
compound has a unique absorption—a
plot of light absorbed versus wavelength—which can be derived from a
solution (see ABSORBANCE). Such spectra
are used to identify compounds, determine concentrations, and plot reaction
rates.

acentric fragment A fragment of a
chromosome not containing a CENTRO-
MERE. This absence prevents acentric
fragments from segregating at MITOSIS,
and they eventually disappear.

Acetobacter A genus of gram-negative flagella-endowed bacteria that are
acid-tolerant aerobic rods. Because of
their ability to oxidize ethanol to acetic
acid, they are also known as acetic acid
bacteria. *Acetobacter* are found on fruits
and vegetables and can be isolated from
alcoholic beverages. Although used commercially in the production of vinegar,

because of their ability to produce acetic acid they are nuisance organisms in the brewing industry.

Acetobacter aceti   An organism used in the commercial production of vinegar. When introduced into wine or cider containing 10–12% alcohol, the organism converts it to acetic acid. (See ACETOBACTER.)

acetone-butanol fermentation The anaerobic fermentation of glucose by *Clostridium acetobutylicum* to form acetone and butanol as end products. At one time, the production of these commercially important chemicals relied upon bacterial fermentation, but it now uses chemical synthesis.

acetylcholine   A chemical neurotransmitter expelled into the synaptic cleft or space between two nerve cells, which permits transmission of an electrical nerve impulse or action potential from one nerve cell to another by diffusing across the cleft and binding to a cell membrane receptor.

acetylcholinesterase   An enzyme present in the synaptic cleft (the space between two nerve cells) that hydrolyzes or destroys the unbound neurotransmitter ACETYLCHOLINE after it has diffused through the cleft. This enzyme is required to restore the synaptic cleft to a state ready to receive the next nerve impulse.

acid blobs   Certain sequences of amino acids in a protein that bind to a transcriptional regulatory protein and thus serve to activate TRANSCRIPTION.

acidic amino acids   Amino acids negatively charged at pH 7.0, namely aspartic acid (aspartate) and glutamic acid (glutamate). They contain a second carboxyl group that is ionized under physiological conditions.

acidophile   Classification of microorganisms that describes the ability or necessity of certain species to exist in an acidic environment. These acid-loving organisms can exist at a pH range of 0–5.4, well below the optimum of neutrality for most bacteria. Facultative acidophiles (see AEROBE) can tolerate a range of pH from low to neutral and include most fungi and yeasts. However, obligate acidophiles, including members of the genera *Thiobacillus* and *Sulfolobus,* require low pH for growth; a neutral pH is toxic to them.

acquired immune deficiency syndrome (AIDS)   An infectious disease in humans caused by the human immunodeficiency virus (HIV). The virus attacks the host's immune system, leaving the individual susceptible to many other diseases, including certain rare forms of cancer and opportunistic microbial infections that would otherwise be destroyed in an uninfected individual, and from which the patient dies. The virus is transmitted through exchange of body fluids during sexual contact with an infected individual, sharing of needles among intravenous drug users, transfusion of contaminated blood products (no longer a threat because donated blood can be screened), and from an infected mother to her newborn during delivery. It has not been shown to be transmitted through casual contact with infected individuals.

acridine orange   One of a group of chemical mutagens known as acridines, the size of which is the same as a purine-pyrimidine base pair. Hence, they can insert (intercalate) into the helix between two adjacent base pairs. When DNA containing an intercalated acridine is replicated, a base pair may be added or deleted, disrupting the CODON reading frame in the newly synthesized strand. Such a mutation is a frameshift mutation.

acrosome (process, reaction, vesicle)   A vesicle or membrane-bound compartment covering the sperm head and that contains lytic enzymes (see

LYSE). The major enzyme found in the mammalian sperm acrosome is hyaluronidase, which promotes the digestion of the tough outer coat of the egg and allows penetration of the sperm.

acrylamide A substance that can polymerize and form a slab gel when poured into a mold in its molten state. It is used as a semisolid support medium and immersed in a conductive buffer through which an electric current is passed. When solutions containing heterogeneous mixtures of nucleic acid fragments or mixtures of proteins are placed into slots in the gel and subjected to the current, the mixtures may be separated into distinct collections of homogeneous molecules located in different regions of the gel, based on their size or molecular weight.

actin One of two major proteins responsible for muscle contraction. Actin and myosin are found in smooth and striated muscle. Actin monomers together with the proteins troponin and tropomyosin can polymerize to form long thin filaments that, together with myosin filaments, can shorten in the presence of ATP. Actin also plays a role in the shape and structure of cells.

Actinomyces A genus of anaerobic gram-positive rods often found in the mouth and throat. They occasionally display a branched filamentous morphology. Many, such as *A. israelii,* are human pathogens.

actinomycin D An antibiotic produced by *Streptomyces parvullus* that inhibits RNA transcription in eucaryotes. It blocks the action of RNA polymerase I, which synthesizes ribosomal RNA. Actinomycin D can intercalate into DNA between G-C base pairs and block elongation of RNA. Although toxic, it is sometimes used in conjunction with other drugs as a chemotherapeutic agent, due to its antitumor properties.

action potential A sequential wave of depolarization and repolarization across the membrane of a nerve cell (neuron) in response to a stimulus. Depolarization is a reversal in the distribution of charge between the inside and outside of the neuron membrane. Also called a nerve impulse.

activated sludge process A secondary sewage treatment process where biological processing of the sewage by microbial activity is the main method of treatment. Sewage previously treated in settling tanks is aerated in large tanks to encourage growth of microorganisms that oxidize dissolved organics to carbon dioxide and water. Bacteria, yeasts, molds, and protozoans are used. This process proves effective in reducing intestinal pathogens in sewage while encouraging growth of nonpathogens. After activated sludge has been produced, additional processing is required, including anaerobic digestion, filtering, and chlorination.

activation energy The energy required for a chemical reaction to proceed. In biological systems enzymes lower the activation energy, allowing chemical reactions to occur faster under physiological conditions.

active site The region of an enzyme containing a special binding site for substrate(s). This site is uniquely shaped for the exclusive binding of the particular substrate molecule(s) and is the site for the catalytic activity of the enzyme. Three-dimensional folding of the enzyme brings distal amino acids in the polypeptide into close proximity, thus forming the active site at the protein surface.

active transport Transport of ions and metabolites by cells or cellular compartments through cell membranes against a concentration gradient. It requires energy in the form of ATP hydrolysis (see ADENOSINE TRIPHOSPHATE). One example found in all animal cells is the active transport of Na^+ out of cells and K^+ into cells, a system known as the sodium–potassium pump. The energy is provided by a specific ATPase located in the plasma membrane, and the system is responsible for generating and maintaining the electric potential or voltage gradient across the cell membrane.

acycloguanosine (acyclovir) An antiviral antibiotic used to treat herpesvirus infections. It is a derivative of the normal nucleoside, guanosine, in which the sugar, ribose, has been replaced by an ether chain. Acycloguanosine is an inhibitor of viral DNA synthesis. (See HERPES SIMPLEX VIRUS (HSV).)

acyl carrier protein (ACP) A small protein involved in the synthesis of fatty acids. First isolated from *E. coli* bacteria by Vagelos, it was found to be a 77-amino acid polypeptide chain capable of binding six other enzymes required for fatty acid synthesis.

adaptation (1) *sensory* A progressive decrease in the number of impulses passing over a sensory neuron even when there is continuous or repetitive sensory stimulation to the sense organ involved. Sensory adaptation provides an organism's sense organs with a way to deal with constant bombardment by useless information in an environment and the ability to respond to only the appropriate stimuli. (2) *evolutionary* A genetic change in a population of organisms that arises as a result of random chance, involving structures or behaviors that will enable that group and its offspring to be better suited to their environment.

adaptive enzymes Enzymes produced by microbes only when their substrates are present. When not needed, these enzymes are not produced, unlike constitutive enzymes, which are always produced.

adaptor molecules A term used to describe transfer RNA (tRNA) due to its role during translation of messenger RNA (mRNA). Several properties of the tRNA molecule enable it to act as an adaptor molecule. The highly specific nature of tRNA–amino acid binding, the complementary base pairing of the tRNA ANTICODON with a specific CODON in the message, and its ability to carry its designated amino acid to the mRNA template in the ribosome are all factors that allow information in the message to be translated to a polypeptide.

adenine One of four major bases found in nucleic acids. Adenine and guanine are purines; cytosine, thymine, and uracil are pyrimidines. These nitrogenous bases are a component of the basic building blocks of nucleic acids called nucleotides. Within the DNA double helix, adenine forms a double hydrogen bond with thymine.

adenoma A benign tumor of glandular tissue.

adenomatous polyposis coli (APC) syndrome A GENETIC DISEASE characterized by the development of benign polyps in the colon, a condition that frequently precedes development of malignant colon cancer. The genetic locus of APC has been shown to reside on human chromosome 5. Research aimed at mapping and then cloning the causative gene(s) via chromosome walking is currently under way.

adenosine A nucleoside containing the pentose sugar ribose and the purine base adenine. When a phosphate group is attached to the 5′-carbon in the ribose, the nucleoside becomes a nucleotide, a basic building block of nucleic acids.

adenosine diphosphate (ADP) One of two products of hydrolysis or enzymatic cleavage of the terminal phosphate group of adenosine triphosphate (ATP), the other being inorganic phosphate, to provide energy for cells to do work.

adenosine monophosphate (AMP) A product of the hydrolysis or enzymatic cleavage of the two terminal phosphate groups of ADENOSINE TRIPHOSPHATE

(ATP). The other product produced during this reaction is inorganic pyrophosphate. ATP is cleaved to provide energy for cells to do work.

adenosine triphosphate (ATP) The single most important energy source in biological systems, it is the energy currency of the cell. All cells do work that requires energy. Work can be mechanical, biosynthetic, active transport of molecules into and out of cells and cellular compartments (see ACTIVE TRANSPORT). All of these processes require hydrolysis of ATP or breaking of phosphate bonds in the energy-rich ATP molecule. ATP is composed of adenine, which is linked to the 5-carbon sugar ribose. In addition, three phosphate groups are linearly linked to the ribose. The two terminal phosphate groups possess high-energy bonds that, when cleaved, provide energy for the cell.

adenovirus A DNA-containing virus with an outer protein coat shaped like an icosahedron. There are more than 40 different types of adenoviruses, some of which are among the many viruses responsible for the common cold.

adenylate cyclase An enzyme that catalyzes the synthesis of CYCLIC AMP from ATP. (See ADENOSINE MONOPHOSPHATE, CYCLIC AMP, ADENOSINE TRIPHOSPHATE.)

adhering junction A type of cell-cell junction that is a highly specialized region of the cell's surfaces, commonly found in tissues subjected to mechanical stress, such as the skin. The junction provides very tight contact between adjacent cells and enables groups of cells to function as a unit in tissues. Also called a DESMOSOME.

adhesion plaque Specialized region of the plasma membrane involved in the adherence of cells to solid surfaces. Bundles of actin microfilaments called stress fibers attach to the plasma membrane in adhesion plaques and are anchored by the protein vinculin. When cells are transformed to a cancerous state, the adhesion plaques become disordered and cells lose their ability to adhere properly (loss of anchorage dependence), thus contributing to metastasis.

A-DNA One of several forms assumed by a double helix under different conditions in vivo or in vitro. Its molecular characteristics differ from the more common B form, which appears to dominate under physiological conditions: it is stable in a less humid milieu and is both the form of a DNA-RNA hybrid helix and the conformation assumed by regions of double-stranded RNA. Although a right-handed helix, A-DNA is more compact than B-DNA.

adoptive immunity The transfer of immunity to allografts (see ALLOGRAFT IMMUNITY) to an animal that was previously tolerant of such allografts by injection of lymphocytes from an animal that is immune to allografts into the tolerant animal.

adrenergic Pertaining to the general class of neurons that uses catecholamines (adrenaline, dopamine, and noradrenaline) as NEUROTRANSMITTERS.

adrenocorticotrophic hormone (ACTH) A hormone secreted by the anterior lobe of the pituitary gland that controls the production and secretion of adrenal cortex hormones. It is a tropic hormone because it regulates the activity of other hormones. It in turn is regulated by a regulating factor produced by the hypothalamus. Under stress, the anterior pituitary secretes ACTH, which in turn stimulates the adrenal cortex to secrete glucocorticoids, which promote gluconeogenesis (see GLUCOGENESIS), a metabolic process involving synthesis of glucose from various noncarbohydrate metabolites in the cell.

adsorption A step in the replication of bacterial viruses where the virus attaches to a specific receptor located on the outer surface of the cell. The receptor is complementary to the attachment site on the virus. The specificity of a virus for a particular host or a few hosts is a result of the fact that the virus can only adsorb to species of bacterial cells that make appropriate receptors. Following adsorption, the PHAGE GENOME penetrates the cell where it is replicated, transcribed, and translated, and viral components self-assemble into new viral particles. This process is followed by cell lysis, or the bursting open of the cell, and the release of the newly synthesized VIRIONS, which can number 50–200, depending on the virus.

aeration The process whereby small bubbles of air or oxygen are introduced to liquid cultures of bacteria with agitation or stirring to ensure that the cells are receiving a continuous and adequate supply of molecular oxygen. Aeration techniques are applied to growth of microbes in industrial fermentors of large-volume capacities and in ordinary flasks as well on a gyrating platform in an incubator or water bath.

aerobe A microorganism requiring free oxygen for growth. During respiration oxygen is the final electron acceptor in the electron transport chain in an aerobe. Obligate aerobes die without oxygen. Microaerophiles thrive in the presence of low amounts of oxygen, and facultative anaerobes normally use oxygen but can switch to an anaerobic metabolism when oxygen is depleted. Aerobes are more efficient at producing energy than organisms that do not use oxygen.

aerosol A mist or cloud of water droplets suspended in air that can carry airborne pathogens and provide a vehicle for transmission. Aerosols may be formed in the environment in numerous ways, such as coughing, sneezing, splashing of falling raindrops, and spray from breaking waves.

aerotolerant Describing anaerobes that do not use oxygen during metabolism but, unlike obligate anaerobes, can survive in its presence (see ANAEROBE). Members of the genus *Lactobacillus* represent examples of aerotolerant microorganisms.

afferent Refers to the direction in which a nerve impulse is moving toward the central nervous system. Afferent neurons are nerve cells carrying impulses from sensory organs (skin, tongue, etc.) toward the central nervous system (brain and spinal cord), whereas efferent neurons carry impulses from the central nervous system to effector organs (e.g., muscle).

affinity chromatography A technique for purification of a protein or other biomolecule, from a mixture or crude sample, based on the natural affinity of the biomolecule for a particular HAPTEN. In affinity chromatography the crude mixture is passed over a column containing a solid, inert matrix, such as an agarose, to which the hapten is covalently bound.

aflatoxin A highly toxic chemical in a class of compounds called mycotoxins, produced by molds. Aflatoxin is produced by *Aspergillus flavus,* which grows on grains and has been found to contaminate many foodstuffs, including beans, cereals, and peanuts. Aflatoxin has been shown to be one of the most potent liver carcinogens in existence.

African sleeping sickness A disease, also known as African trypanosomiasis, affecting humans and other mammals in central Africa. It is spread by the tsetse fly, which is host to the parasitic protozoan the trypanosome *(Trypanosoma brucei gambiense* and *T. brucei rhodesiense),* the causative agent. After being bitten by the fly, the trypanosome enters the victim's bloodstream. Without treatment the disease is nearly always fatal because the trypanosome enters the central nervous system and coma ensues. Trypanosomes can evade the host's immune system because they repeatedly change their coat proteins against which the host makes antibodies. This pathogen is the subject of intense study due to its devastating effects on humans and livestock and to the unusual characterisitics of its molecular biology.

agar   A complex POLYSACCHARIDE made by the red marine alga *Gelidium,* used to thicken or solidify bacterial culture media and certain foods. It was first used for culturing microorganisms, for which it is well suited, by the wife of Walter Hesse, a German microbiologist in the late 1800s. Very few microorganisms can degrade or digest agar, so it remains solidified in their presence and at high enough temperatures (close to 100°C) to be able to incubate almost all microbes. When in a molten state, it will solidify below 42°C, but can be kept as a liquid for long periods if incubated at 50°C and above. Agar can be poured into tubes, flasks, petri plates, or any other support and placed in any position (such as slanted or straight), to solidify, to shape the surface in order to maximize or minimize oxygen availability and surface area. Most solid media are 1.5% agar.

agarose   A cross-linked polysaccharide isolated from red algae and used to pour gels used for agarose gel electrophoresis.

agarose gel electrophoresis   A procedure that uses agarose gels to separate molecules in solutions of nucleic acids or solutions of proteins according to their size. In this way, molecular weights can be determined, or certain specific species of molecules can be isolated and purified. Size ranges of fragments that can be separated are determined in part by the percentage of agarose in the gel. The gels are immersed in a chamber containing a buffer that can conduct an electric current across the gel. The samples are loaded into slots in the top of the gel, a dye is added, and the current is turned on.

agglutination   The clumping of cells to one another caused by the binding of agglutinin molecules to the cell surface so that one or more cells are linked to one another by an agglutinin bridge. (See HEMAGGLUTINATION.)

Agrobacterium   A genus of gram-negative aerobic bacteria that live in soil and cause crown gall disease in broad-leafed plants, evidenced by the growth of tumors on the trunks and sometimes the roots. The pathogenicity of the organisms is due to the presence of a bacterial PLASMID, called the Ti plasmid, that can be transferred to the plant cells from the bacteria. The plasmid contains genes that direct the plant cells to make nutrients useful for the bacteria and gene products that interfere with normal plant cell growth and division. Microbial geneticists and molecular biologists are intensely studying *A. tumefaciens* with the hopes of being able to use this organism to transfer useful genes into crop plants, which can be accomplished by using the Ti plasmid as a vehicle to transfer genes, such as those involved in nitrogen fixation, into crop plants after the plasmid has been genetically engineered to eliminate its pathogenicity.

Agrobacterium tumefaciens   A bacterium characterized by a special spore-containing sac. (See AGROBACTERIUM.)

AIDS related complex (ARC)   A series of symptoms related to an active human immunodeficiency virus (HIV) infection, including general malaise, night sweats, dementia, wasting, and opportunistic diseases associated with immunodeficiency, such as Karposi's sarcoma (a rare form of skin cancer), pneumonia (generally caused by *Pneumocystis caranii*), and retinitis (caused by cytomegalovirus).

alanine   One of 20 amino acids incorporated into POLYPEPTIDES. Alanine has

An Alarmone

diadenosine 5',5'''-P^1,P^4-tetraphosphate

an aliphatic uncharged R group at pH 7.0 consisting of a methyl (CH_3) group as its side chain. Of all the amino acids with aliphatic side groups, it is the least hydrophobic.

alarmones   Unusual dinucleotides containing multiple phosphate groups produced by bacteria under conditions of stress, such as exposure to oxidative agents (e.g., hydrogen peroxide, and may act in a hormone-like fashion to regulate bacterial metabolism under such conditions.

albumin   The most abundant human blood plasma protein. It is a heat-coaguable, water-soluble globular protein found in egg white, blood plasma (50% of protein content of human plasma), and various other animal and vegetable tissues. Bovine serum albumen is often used in reaction mixtures and storage tubes to stabilize enzymes.

alcohol   An organic compound containing one or more hydroxyl groups (—OH). Also the common name for ethanol. Other industrially produced alcohols include methanol and propanol.

alcohol dehydrogenase   An enzyme responsible for the last step in alcoholic fermentation by yeast, which produces the alcohol in alcoholic beverages. The enzyme converts acetaldehyde to ethanol.

alcohol fuel   An energy source produced by bioconversion, in which organic waste material is converted to fuel by microorganisms. Examples are gasohol (90% gasoline and 10% ethanol), an alternative fuel for automobiles, and a by-product of the anaerobic treatment of sewage and an alternative to fossil fuels and natural gas.

aldose   A group of monosaccharides containing an aldehyde group (—CHO).

aldosterone   A hormone secreted by the adrenal cortex and classified as a mineralocorticoid, it acts mainly on the kidneys to control the water and electrolyte balance in the body. It ensures the retention of sodium ions and water by causing their reabsorption into the blood before urine excretion and also causes the excretion of potassium ions in the urine.

algae   Photosynthetic eucaryotic organisms classified as either protista or plants according to their morphology. Some algae exist as single-celled organisms, and some are multicellular. They are usually aquatic, occupying both marine and freshwater environments. Dinoflagellates and diatoms are free floating, and red and brown algae require a solid substrate to which they attach. Algae are further classified by their photosynthetic pigments, hence the names brown, red, green, and blue-green algae. Many are of industrial importance, providing thickeners for foods and bacterial culture media. (See AGAR.)

alkaline phosphatase An enzyme isolated from bacteria, commonly used in genetic engineering experiments to remove the 5'-phosphate groups from the ends of DNA or RNA molecules.

alkaloids A class of over 3000 compounds containing nitrogen that are produced by plants but that exert potent physiological effects on animals. They are synthesized from aromatic amino acid precursors such as tyrosine, tryptophan, and phenylalanine. Examples are morphine, cocaine, nicotine, codeine, and colchicine.

alkalophiles (alkalinophiles) Microorganisms that flourish in basic environments (base loving), pH 7–12. They include *Vibrio cholerae*, the causative agent of cholera, whose optimal pH is 9.0, and the soil bacterium *Agrobacterium,* whose optimal pH is 12.

alkaptonuria The first human genetic disease identified as such when it was found to follow the laws of Mendelian inheritance. Also known as "dark urine disease," it was studied by Garrod, and in 1902 was recognized to be inherited as a recessive trait. Later its biochemical nature was also uncovered. The disease is characterized by a deposit of dark pigment in connective tissue and in the urine after exposure to air. Later stages result in severe forms of arthritis and possibly death due to blockages in the arteries and heart valves. One in a million people is born with this disease, which is caused by a deficiency in the enzyme homogentisic acid deoxygenase,

resulting in the accumulation of homogentisic acid in the urine.

alkylating agent A mutagen that adds alkyl groups, such as the methyl group ($-CH_3$) and the ethyl group ($-CH_2CH_3$), to bases in DNA. One such mutagen is ethylmethane sulfonate (EMS), which can alkylate either thymine or guanine residues and cause them to mispair during DNA replication, resulting in transition-type mutations in DNA.

allele One of several alternative forms of the same gene. (A single gene can have as few as 1 or as many as 100 different alleles.) Alleles are differences in the base sequence of a single gene among individuals in a population or on the two homologous chromosomes in one individual. They are the cause of genetic variation or different expressions of a trait in a population of organisms.

allergen A type of antigen that causes an allergic response. Allergens appear to interact with IgE-type antibody molecules located on the surfaces of mast cells.

allograft immunity The state of the immune system in which grafted tissue originating from a genetically dissimilar animal provokes attack by the immune system of the host animal (i.e., graft rejection).

allolactose A derivative of lactose and the true inducer of the lactose operon in bacteria. Inside the cell, lactose is converted to allolactose, which in turn activates the three structural genes involved in the utilization of lactose as a carbon source. When lactose is present in the medium, the genes required for its breakdown are active; when it is not present, they are shut off (See LAC OPERON.)

allopolyploid A hybrid organism, usually a plant, bred from two closely related species and containing one or more extra full sets of chromosomes. For example, if each parent has two sets of chromosomes, the allopolyploid offspring, instead of having the normal two sets, may have four. The hybrid contains genetic information different from either parent.

allopurinol   A derivative of the purine base, hypoxanthine, used to treat gout. As an inhibitor of the enzyme xanthine oxidase, allopurinol acts by preventing the accumulation of URIC ACID.

all-or-nothing phenomenon   Refers to the condition that a nerve cell must receive its threshold level of stimulation in order to respond and start an action potential. A nerve will either fire an impulse or not fire at all, if the stimulus is below threshold. There is no such thing as a weak response to a weak stimulus. (See ACTION POTENTIAL.)

allosteric enzymes   Enzymes with many subunits and ACTIVE SITES. They display substrate-induced conformational changes and have different roles or functions in their different conformations. They play an important role in the regulation or metabolic pathways and gene expression.

Alu **elements**   A family of related DNA sequences widely and randomly dispersed through mammalian genomes (about 600,000 copies in the human genome). They are about 300 base pairs in length and are classified as moderately repetitive DNA sequences, at the ends of which is a cleavage site for the restriction enzyme *Alu*. Their purpose, if any, in the genome is not known.

α-**amanitin**   An antibiotic derived from the *Amanita* mushroom that acts by inhibiting the action of RNA polymerase.

amber codon   The CODON UAG, one of three codons that does not code for an amino acid but represents a stop signal.

amber mutation   A type of genetic mutation in a class called nonsense mutations. An amber mutation arises when a three-base-pair sequence in DNA, called a CODON, such as UUG, coding for a specific amino acid, mutates to a UAG codon, which does not code for any amino acid. UAG is a termination codon because it causes the termination of protein synthesis. Any mutation that results in a UAG termination codon is called an amber mutation. Opal and ochre mutations are also nonsense mutations.

amber suppressor   Mutations in the ANTICODON of several different tRNA molecules (see TRANSFER RNA) that allow these mutated tRNAs to recognize the AMBER MUTATION UAG. Ordinarily a UAG CODON in a message signals the termination of translation, but a tRNA with an amber suppressor mutation has an anticodon that is complementary (CUA) to the termination codon. Therefore, at that site in the growing polypeptide chain, it can insert the amino acid that it is carrying and avert chain termination.

ambient   The physical conditions in an organism's surrounding environment. Microorganisms that live in the human gut, for example, have an ambient temperature of 37°C. Organisms that exist in dust particles in a room have an ambient temperature of about 23°C, or room temperature. Ambient conditions also can include atmospheric pressure, humidity, oxygen levels, and other physical parameters.

Ames test   A method for screening potential mutagens and carcinogens, developed by Bruce Ames in the early 1970s, that drastically reduced the time and expense involved in animal testing. The Ames test requires the use of bacteria to determine the possible mutagenic potential of a chemical. It relies on the principle that the chemical structure and properties of DNA are universal. In addition, the mechanisms for toxicity in a bacterium mimic those of a mammal if the appropriate liver enzymes are provided to process the chemical in the same way a mammal would. A chemical that causes mutations in bacteria would likely do the same in a mammal, and since 90% of all known carcinogens are mutagens, a chemical found to be a mutagen in the Ames test would be a suspected carcinogen.

amide   The product of the reaction between an amine compound, a molecule with an NH_2 group, and a carboxylic acid, a molecule with a COOH group. PEPTIDE BONDS ($-CONH-$) link amino acids in proteins and are a type of amide bond between two amino acids. The amino group of one amino acid is liked to the carboxyl group of the next amino acid.

amine   Compound that contains an AMINO GROUP (NH_2). AMINO ACIDS are amines and also carboxylic acids.

amino acids   The building blocks of proteins, containing a free carboxyl group (COOH), a free amino group (NH_2), a hydrogen atom, and a variable side group (R), attached to a single carbon. (One exception to this is proline, whose amino group is part of a cyclic structure.) Physical and chemical properties vary among R groups. However, several classifications put certain R groups in the same category because they share similar properties, namely acidic, basic, aliphatic, aromatic, and hydroxyl-containing or sulfur-containing amino side groups. Twenty different amino acids are found in proteins (See APPENDIX IV.)

aminoacyl-tRNA   A   tRNA   (see TRANSFER RNA) that is carrying its specified amino acid; also called a charged tRNA. The specificity of charging of tRNA molecules is carried out by 20 different enzymes called aminoacyl tRNA synthetases. Each of the 20 amino acids incorporated into proteins is the substrate of one of the enzymes. In addition, each enzyme recognizes the appropriate tRNA(s) to charge and each of the charged tRNAs contains an ANTICODON appropriate to the amino acid that it carries.

aminoglycoside antibiotics   A group of antibiotics that act to kill a broad range of bacteria by interfering with protein synthesis at the bacterial RIBOSOME. They are produced naturally by members of the soil-dwelling genus *Streptomyces* and include streptomycin and kanamycin.

aminoglycoside-3'-phosphotransferase (APH)   A bacterial gene that codes for an enzyme that confers resistance to the antibiotic neomycin. The APH gene is commonly used as a selectable marker in TRANSFECTION experiments in that cells without the gene can be eliminated from a population by exposure to neomycin. (See NEGATIVE SELECTION.)

amino group   The $-NH_2$ group in a molecule. The presence of an amino group is the defining characteristic of the group of organic compounds known as amines.

6-aminopenicillic acid   A chemical structure found in the different natural and semisynthetic penicillins. This com-

Aminoacyl tRNA (charged tRNA)

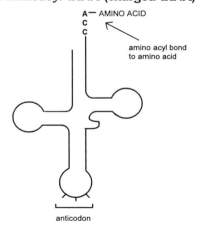

anticodon

mon nucleus of the penicillins contains the beta-lactam ring structure. In addition to the common core, 6-aminopenicillic acid, all penicillins contain a variable side group that distinguishes them from each other.

2-aminopurine   A purine derivative that is a potent mutagenic agent (see MUTAGENESIS) because it becomes incorporated into DNA in place of adenine. As a result, it induces mistakes in DNA during DNA replication.

amino sugars   Derivatives of simple sugars that have been modified to form amines, because of the addition of an NH_2 group in place of the hydroxyl group normally found at carbon 2. Two commonly found amino sugars are D-glucosamine, a major component of chitin, the outer hard covering of insects and D-galactosamine, found in cartilage.

amino terminal   An end of a polypeptide chain consisting of an unreacted amino group. Also called the N-terminus, and the other end is called the carboxyl or C-terminus.

amphibolic   A metabolic pathway that can degrade metabolites and synthesize them—that is, a pathway that is both catalytic and anabolic. These pathways allow breakdown products of one pathway to be used as substrates in the synthesis of a compound in another pathway.

amphipathic compound   A compound containing both polar and nonpolar groups. Polar groups are soluble in water (hydrophilic), and nonpolar groups are not (hydrophobic). In water or aqueous environments, amphipathic compounds form micelles, or small vesicles with polar regions in contact with water and nonpolar regions sequestered in the center of the micelle away from water. Fatty acids are amphipathic. Amphipathic molecules are responsible for the properties of biological membranes.

amphoteric   Description of a substance that has both acidic and basic groups and has properties of acids and bases.

ampicillin   A semisynthetic form of the antibiotic penicillin. The addition of an amino group made it a more versatile drug. All penicillin drugs and their derivatives kill bacteria by inhibiting the formation of bacterial cell walls.

amylose   A starch made up of a long unbranched chain of glucose. A polymer of monosaccharides is called a polysaccharide. Amylose is the principal storage starch of plants.

anabolic   A type of metabolic pathway in which complex molecules are synthesized from smaller precursors, usually in a series of steps. It is the type of metabolism that builds molecules, as opposed to catabolic metabolism, which is degradative. Energy is usually required for anabolic metabolism. Examples are the synthesis of polypeptides from amino acids or the synthesis of nucleic acids from nucleotides. (See CATABOLISM.)

anaerobe   A microorganism that does not or cannot use oxygen during RESPIRATION. Obligate anaerobes such as the genus *Clostridium* die in the presence of oxygen. Others, such as *E. coli,* are classified as facultative anaerobes because they use oxygen when present but can switch to anaerobic RESPIRATION in its absence.

anaphase   A stage during MITOSIS, or cell division, where chromosomes split at the CENTROMERE and the resulting chromatids move to opposite ends of the cell. During MEIOSIS, or reduction division, there are two anaphase stages. During anaphase I, homologous pairs of chromosomes separate from each other with their centromeres intact and move to opposite ends of the cell. Anaphase II resembles the anaphase stage in mitosis, as described above. (See CENTRO-MERE, MEIOSIS, MITOSIS.)

anaphylotoxin A substance released by the body as part of an immunological response to the presence of a foreign antigen. Anaphylotoxins stimulate the release of histamines, which cause inflammation in tissues.

androgens A group of male sex hormones responsible for the development and the maintenance of masculine features and organs. Testosterone is an androgen.

angstrom (Å; AN) A unit of measurement usually used for wavelengths or cellular structures. 1 Å$=10^{-10}$ meter, or 10^{-6} millimeter, or 10^{-4} (0.0001) micrometer, or 0.1 nanometer.

anion A negatively charged ion.

anneal Complementary single strands of DNA, or DNA and RNA, forming hydrogen bonds between complementary base pairs to form double-stranded DNA or DNA-RNA hybrids.

antennapedia complex A genetic locus in the homeotic box defined by mutations that cause developmental defects in the thoracic and head segments of the fruit fly, *Drosophila melanogaster*. (See HOMEOBOX.)

antibiotic A substance, usually made by a microorganism, that inhibits the growth of another microorganism or kills it (e.g., penicillin). Many synthetic antibiotics are derivatives of naturally occurring antibiotics available for medicinal or research purposes.

antibiotic resistance Microorganisms may have a natural resistance or develop resistance to an antibiotic, in which the drug is not effective in inhibiting growth of the microorganism or killing it.

antibiotic resistance genes Genes that confer antibiotic resistance to a microorganism. Examples are genes that (1) encode enzymes that destroy the antibiotic, (2) code for the target of the antibiotic but become mutated, so the target no longer responds to the drug, or (3) encode proteins preventing the antibiotic from being taken up by the microorganism.

antibodies Proteins that circulate in the bloodstream and bind to foreign invading substances (antigens) (e.g., bacteria, toxins, certain viruses) with a great deal of specificity. Antibodies, such as immunoglobulins, are the mediators of the immune response to soluble antigens.

antibody-producing cell An activated B lymphocyte or plasma cell that secretes antibodies. Each plasma cell secretes an antibody with specificity for one antigen.

anticoagulant A chemical substance that prevents the coagulation of blood.

anticodon A three-nucleotide base-pair sequence that is antiparallel and complementary to a codon. The anticodon is found on a tRNA and interacts with a specific codon on the mRNA so that an amino acid will be placed in the correct position according to the mRNA during translation or protein synthesis.

antidiuretic A chemical substance that counteracts a diuretic.

antifungicide A substance or drug that kills fungi.

antigen A substance that stimulates the production of specific neutralizing antibodies in an immune response. Any chemical substance, usually protein, that interacts with an antibody.

antigenic determinant A small portion of the antigen that determines the specificity of the antigen-antibody reaction.

antigenic variation   A sequential change in the structure of an antigen of microorganisms and viruses so that these antigens will not be recognized by antibodies already produced in the host. The disease relapses in fever, which is characterized by cyclic infections by the same bacterium and this is due to the ability of the bacterium to change its antigenic makeup and thus avoid immunity built up by the host. A subtler type of antigenic variation is seen in the antigenic shift (major antigenic change) and antigenic drift (minor antigenic change) in the influenza virus, which results in loss of immunity by populations and influenza pandemics and epidemics every few years.

antigen   processing/presenting cell   Any of a heterogeneous group of cells that bind foreign antigens to their surface and then interact with helper T cells, a process required for T-cell activation. Antigen-presenting cells include dendritic cells in lymphoid tissue, Langerhans cells found in skin, and some MACROPHAGES.

antihelminthic agent   A substance or drug that inhibits the growth of, or kills, helminth (worm) parasites.

antihistamine   A substance or drug that blocks the effects of histamines in the inflammatory process; a drug that relieves allergy symptoms.

antimetabolite   A chemical that inhibits the growth of microorganisms because it blocks the synthesis of some metabolite needed by the microorganism, such as sulfa drugs, which block the synthesis of the vitamin folic acid.

antimutator   A gene that decreases the spontaneous mutation rate of an organism. These genes are usually involved in some DNA repair or metabolism process.

antioncogene   A tumor-suppressor gene, or a gene whose absence is needed for an oncogenic event. Loss or inactivation of a tumor-suppressor gene by either mutation or deletion is believed to be an important event in the development of a tumor.

antiparallel   Refers to the structure of DNA (see DEOXYRIBONUCLEIC ACID). The two strands of complementary DNA are antiparallel; that is, the 5'-end of one strand is paired with the 3'-end of the other, and vice versa. (See POLYNUCLEOTIDE.)

antiparasite   A Substance or chemical that inhibits or kills parasites.

antiport   The transport of two substances, which are coupled but in opposite directions, across the cell membrane.

Antiparallel Strands in Double-stranded DNA

antisense RNA   A strand of RNA (see RIBONUCLEIC ACID) complementary to that of mRNA. An antisense RNA binds to the messenger and prevents synthesis of the protein encoded by the message. Antisense RNA is being explored as a possible therapeutic agent for viral infections and to prevent certain cancer genes from being expressed into proteins.

antisense strand   Of the two DNA strands in a double-stranded DNA molecule, the one not used as the template for RNA synthesis.

antiseptic   Any chemical commonly used to kill microorganisms to prevent infection.

antitermination factor   A protein that prevents the termination of transcription. It is involved in certain mechanisms of gene expression control.

apoenzyme   The protein moiety or part of an enzyme without its cofactor; normally inactive.

apurinic site   A site on the DNA in which a purine is missing, but the phosphodiester sugar backbone is still intact.

apyrimidinic site   A site on the DNA in which a pyrimidine is missing, but the phosphodiester sugar backbone is still intact.

aqueous   Pertaining to water; for example, the aqueous phase after separation with an organic solvent is the water phase.

arabinosyladenine   *(araA)*   An antiviral antibiotic used to treat viral encephalitis. *araA* is a derivative of the normal purine nucleoside adenosine, in which the sugar, ribose, has been replaced with one of its optical isomers, arabinose.

arabinosylcytosine *(araC)*   An antibiotic that acts by blocking DNA syn-

Arabinosylcytosine *(araC)*

thesis. It is a derivative of the normal pyrimidine nucleoside cytosine, in which the sugar, ribose, has been replaced with one of its optical isomers, arabinose.

arachidonic acid   A 20-carbon fatty acid with four double bonds that serves as a precursor for the synthesis of prostaglandins.

archebacteria   A group of bacteria, including those that produce methane from carbon dioxide and hydrogen (methanogens) and those that live in high-salt environments (halophiles), that appear to be very different from and more primitive than other living bacteria. They are believed to have separated very early from other present-day bacteria during evolutionary history.

arginine   An amino acid with side chain $-(CH_2)_3-NH-C=NH$
$$\backslash$$
$$NH_2$$

aromatic   An organic compound containing a benzene or benzene-derived ring.

Arthus reaction   An inflammatory response caused by the production or deposition of antigen-antibody complexes in tissues.

artifact The appearance of a structure in microscopy or an experimental result that is not real but due to experimental procedures used.

Ascomycetes A class of fungi distinguished by an ascus, a saclike structure that produces ascospores, or the sexual spores of the *Ascomycetes.*

ascorbic acid Vitamin C, a dietary lack of which leads to scurvy. Ascorbic acid is a reducing agent that keeps the enzyme prolyl hydrolase in active form. Collagen synthesized in the absence of ascorbic acid is insufficiently hydroxylated, cannot form fibers properly and causes the skin lesions associated with scurvy.

aseptic Without germs; sterile.

asexual reproduction Reproduction in the absence of any sexual process, or the reproduction of a unicellular organism by cell division where a single parent is the sole contributor of genetic information to its offspring.

asparagine An amino acid with side chain $-CH_2-C=O$ with NH_2

aspartic acid An amino acid with side chain $-CH_2-C=O$ with OH

Aspergillus An economically important genus of fungi used in industrial fermentations.

assay A test. In an enzyme assay, an enzyme is tested for activity under specific conditions.

AT content The fraction of total nucleotides in a DNA molecule that are either adenine or thymine nucleotides generally given as a percentage.

AT/GC ratio The ratio of adenine plus thymine base pairs to guanine plus cytosine base pairs in a molecule of DNA.

attenuation (1) A decrease in virulence of a pathogen. (2) A mechanism of gene regulation in bacteria in which availability of certain amino acids will control the expression of genes for their own synthesis by causing premature termination of transcription of the genes involved in the synthesis.

att site A site on the *Escherichia coli* bacterial chromosome that interacts with the BACTERIOPHAGE lambda genome and at which the bacteriophage genome integrates into the bacterial genome resulting in lysogenization of the bacterium. (See LYSE.)

autoclave An apparatus that uses steam under pressure to sterilize materials.

autogenous control Control of gene expression by the gene's product or protein encoded by the gene.

autoimmune The inability to distinguish self from nonself, or a state where the body produces antibodies to its own cells.

autolysin An enzyme that causes cellular self-destruction of the same cells that synthesize it.

autolysis The self-degradation of a cell by release of hydrolytic enzymes of the LYSOSOME. In bacteria, autolysis is brought about by self-destruction of the cell wall by specific enzymes.

autonomic nervous system The part of the nervous system that regulates involuntary responses.

autonomously replicating sequences (ARS) Special nucleotide sequences in the DNA of chromosomes that serve as sites where DNA replication begins.

autonomously replicating sequences (ARS) element Found on yeast

PLASMIDs that are initiation sites for plasmid replication. Plasmids lacking ARS sites will not replicate.

autoradiography A technique that involves using a radioactively labeled compound to localize a reaction to a cell or to study a process, and using photographic film to visualize the location of the label.

autosome A chromosome that is not a sex chromosome.

autotroph An organism that can make its own nutrients from organic carbon compounds or from inorganic carbon in the form of carbon dioxide.

auxin A plant hormone that regulates cell reproduction and cell elongation in certain tissues.

auxotroph A bacterial mutant that can no longer make some required nutrient.

axenic culture Pure culture or the growth of one organism.

axon Extension of a nerve cell that conducts inpulses away from the cell body.

axoneme The structural core of a cilia (see CILIUM) or eukaryotic flagellum made up of nine outer doublets of microtubules and an inner pair of microtubules.

3'-azido-3'-deoxythymidine (AZT) An antiviral antibiotic used to treat HIV infection (the AIDS virus). AZT is a derivative of the normal deoxyribonucleoside thymidine, in which an azide group is attached to the deoxyribose sugar at the 3' position. AZT is an inhibitor of the virus REVERSE TRANSCRIPTASE enzyme, which blocks viral replication at the point where viral RNA is copied into DNA.

Azotobacter A genus of free-living microorganisms capable of biological nitrogen fixation, or the ability to use nitrogen of the atmosphere for synthesis of nitrogen-containing compounds.

B

Bacillus A genus of free-living rod-shaped bacteria that produces extremely resistant spores, insuring the organism's survival under harsh environmental conditions. Some species produce antibiotics.

bacitracin An antibiotic effective against gram-positive bacteria. It inhibits cell-wall synthesis.

backbone (1) The spinal column of a vertebrate organism. An organic molecule, for example, a polypeptide chain, that is a main structural feature of an entire class of compounds; a compound from which an entire class of

structurally related other compounds is derived.

backcross A genetic cross between a heterozygote and one of its parental homozygotes. (See HETEROZYGOUS, HOMOZYGOUS.)

back mutation A mutation that reverts a previous mutation, so the mutant PHENOTYPE is changed back to (nonmutated) wild type.

bacteria A group of single-celled prokaryotic organisms that divide by binary fission, are HAPLOID (or contain one copy of a chromosome), do not possess organelles such as mitochondria and chloroplasts, and do not have a membrane-bound nucleus.

bacterial transformation A genetic transfer process, where cell-free, isolated DNA is taken up by a recipient cell and incorporated into its genome.

bacterial virus A bacteriophage, or a virus that uses a bacterium as its host to reproduce.

bacteriocidal Describing a chemical or drug that can kill bacteria.

bacteriophage A bacterial virus that utilizes bacterial host replicative systems for its own replication, after which the host cell is usually destroyed, releasing progeny bacteriophage. Each bacteriophage particle consists of an icosohedral-shaped head that carries the bacteriophage DNA genome. The bacteriophage attaches to its bacterial host by means of a cylindrical tail with a hollow core. The tail serves as a conduit for injecting DNA into the host.

bacteriophage λ A DNA-containing bacterial virus that infects *Escherichia coli* and has a complex set of regulatory mechanisms governing whether the virus will reproduce itself and LYSE its host or lysogenize its host by integration of its genome into its host's genome. Derivatives of lambda are used as cloning vectors to introduce foreign DNA into *E. coli.*

bacteriophage M13 A single-stranded DNA phage used as a cloning vector to produce single strands of foreign DNA for DNA sequencing and in vitro mutagenesis procedures.

bacteriophage mu A DNA virus capable of transposition, or inserting itself randomly into the genome of its host. This virus is used in the process of inser-

Bacteriophage

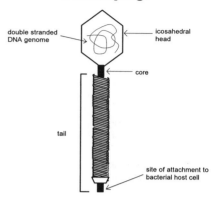

double stranded DNA genome

icosahedral head

core

tail

site of attachment to bacterial host cell

tional mutagenesis. (See INSERTION MUTATION.)

bacteriophage Qβ A single-stranded RNA bacteriophage.

bacteriophage T4 A large DNA virus.

bacteriophage T7 A DNA virus with a very strong PROMOTER that responds to specific T7 RNA polymerase. A number of CLONING VECTORS have been constructed so that foreign DNA is situated next to a T7 promoter so that expression of the gene can be regulated and amplified by addition of T7 RNA polymerase.

bacteriophage φX174 A single-stranded DNA virus used to study the process of DNA replication.

bacteriophage, transducing A phage that acts as a vector in a gene transfer process, by injecting donor bacterial DNA into a recipient upon viral infection.

bacteriorhodopsin A transmembrane protein of the "purple membrane" of *Halobacterium halobium* capable of transporting protons across the bacterial membrane, thereby creating a light-dependent electrochemical proton gradient.

bacteriostatic A chemical or drug that inhibits the growth of bacteria but does not kill them.

bacteroid A group of anaerobic, gram-negative, small rod bacteria.

baculovirus An insect cell virus used as a cloning vector. Proteins made from cloned DNA in baculovirus are glycosylated, a process that does not occur when cloning in bacteria. (See GLYCOSYLATION.)

baffles Structures on the bottom of some culture flasks that increase aeration when growing a culture of organisms in a shaking water bath or incubator.

baker's yeast *Saccharomyces cerevisiae,* a common yeast, or unicellular budding eukaryotic organism that ferments sugars and produces the carbon dioxide used to leaven bread.

Balbiani rings A very large puff indicating transcriptional activity at a site on the POLYTENE CHROMOSOME of certain larval insects.

Baltimore, David (b. 1938) Molecular biologist and virologist who won the Nobel Prize for physiology or medicine in 1975 for the discovery that retroviruses, a group of viruses that have an RNA genome, produce the enzyme REVERSE TRANSCRIPTASE.

BamHI A restriction enzyme (see RESTRICTION ENDONUCLEASE) that recognizes a specific six-base-pair sequence (GGATCC) and cuts in a staggered manner, thus creating single-stranded overhangs (sticky ends) at the cut sites.

Bam islands Repeated sequences of fixed length in a nontranscribed DNA spacer region. The designation comes from the fact that these sequences were first isolated by digestion of the spacer region with the restriction enzyme BamHI.

barophile An organism that grows under conditions of high hydrostatic pressure but cannot grow under normal atmospheric pressure. Such organisms have been isolated from deep seas where the hydrostatic pressure may be 100 atmospheres or more.

barotolerant An organism that can tolerate high hydrostatic pressure.

Barr body A condensed X chromosome seen in the interphase. The genes on it are not expressed, so the chromosome is inactive.

basal body The organelle, embedded at the periphery of the cell, that serves as the base of the cell's locomotive appendages, the cilia and flagella.

basal lamina The thin layer that underlies epithelial cells and consists of various extracellular matrix proteins including laminin and collagen. The thin membrane surrounding the ovarian follicle is also referred to as a basal lamina.

base (1) A substance that decreases the concentration of H^+ ions in solution, or an alkaline substance. (2) A purine or pyrimidine found in nucleic acids.

base analog A purine or pyrimidine base other than the ones normally found in nucleic acids.

base pair (bp) Complementary relationships between purine and pyrimidine molecules that allow adenine to form two hydrogen bonds with thymine or uracil and guanosine to form three hydrogen bonds with cytosine. Base pairing enables nucleic acids to recognize each other and plays an important role in reactions involving nucleic acids, such

Base Pairing in DNA

as DNA replication, transcription, and translation.

base substitution A type of mutation in which one base or base pair is different in the mutant than in the wild type.

basophile An organism that lives in alkaline environments.

batch culture Growth of microorganisms in a closed system under prescribed conditions of medium, temperature, and aeration.

B cells   See B-LYMPHOCYTES.

Beadle, George W. (1903–1991) A geneticist who, in collaboration with Edward Tatum, showed that genes control enzyme production. Beadle and Tatum shared the 1958 Nobel Prize for physiology or medicine with Joshua Lederberg.

Beer–Lambert law The equation stating that the molar concentration of a substance is proportional to how much light of a certain wavelength is absorbed by a solution of the substance: $A=ECL$, where A=absorbance at a given wavelength, E=molar extinction coefficient, C=molar solution concentration, and L= length of light path.

Bence–Jones protein Part of an antibody molecule (the light chain) found in the urine of individuals who have multiple myeloma, a tumor of the bone marrow. These fragments were instrumental in determining the structure of the antibody.

benign Referring to a tumor that does not proliferate and does not invade surrounding tissues.

Berg, Paul (b. 1926) A biochemist who determined that fatty acid oxidation occurs in two steps.

BglII A restriction enzyme (see RESTRICTION ENDONUCLEASE) that recognizes a specific six-base-pair sequence (AGATCT) and cuts the DNA in a staggered manner, creating single-stranded overhangs at the cut site.

bicoid genes A group of genes coding for proteins that play a determining role in the development of the head and thorax in the embryo of the fruit fly, *Drosophila melanogaster*.

bidirectional replication Replication of a DNA molecule by two replication forks moving in opposite directions from a single initiation point.

binary fission Division of one cell into two after replication of the DNA.

biochemical oxygen demand (BOD) A measure of the amount of oxygen consumed in biological processes that break down organic matter in water. A measure of the organic pollutant load.

biochemistry The chemistry of biological systems and processes.

biodegradation Breakdown of chemicals and biological substances by microorganisms.

bioenergetics The field covering thermodynamic principles applied to biological systems.

biomass The total mass of living matter present on the Earth.

biosynthesis The synthesis of molecules in biological systems. These syntheses are carried out in small discrete steps, each step catalyzed by an enzyme, and are energy requiring, usually involving ATP or GTP as energy sources.

biotin A prosthetic group that carries activated carbon dioxide and is bound to the enzyme pyruvate decarboxylase, important because it replenishes one of the intermediates of the KREBS CYCLE.

biotin labeling (biotinylation) A nonradioactive labeling system in which biotin is covalently linked to a nucleic acid.

biphasic growth curve The growth curve of a microorganism characterized by two exponential growth phases separated by a stationary phase. Such a growth curve is produced by culturing the organisms on two carbon sources, in which one carbon source is in a limiting concentration and must be used up before the second carbon source can be utilized.

bithorax A genetic locus in the homeotic box defined by mutations that cause developmental defects in the thorax region of the fruit fly, *Drosophila melanogaster*. (See HOMEOBOX.)

bivalent A synapsed pair (see SYNAPSIS) of homologous chromosomes found in prophase I and metaphase I of MEIOSIS (see MITOSIS); also known as a tetrad.

blastocyst The particular form of the blastula in mammalian development, characterized by a thickened shell of cells at one end while elsewhere a single cell layer surrounds a fluid-filled cavity (the blastocoele).

blastula The stage in animal development in which a ball of cells is formed from the cleavage of cells of the zygote.

blood agar A culture medium in which animal blood, usually rabbit or horse, is added to provide nutrients or to be used diagnostically for hemolysins secreted by certain strains of bacteria.

blood groups See ABO BLOOD GROUP.

blot A nylon or nitrocellulose membrane onto which nucleic acids are transferred for the purpose of hybridization. (See SOUTHERN BLOT HYBRIDIZATION.)

blotting, capillary diffusion A procedure that transfers nucleic acid from a gel to a nylon or nitrocellulose membrane by capillary diffusion—that is, movement of water through the gel and through the membrane, which results in depositing and trapping the nucleic acid on the membrane as the water moves through.

blotting, electrophoretic A variant of capillary diffusion blotting that uses an electrical field to facilitate the transfer of the nucleic acid to the membrane.

blunt-end DNA Both strands of DNA at one end are even; that is, there are no single-stranded overhangs. Often used in reference to restriction enzymes that cut the DNA at the same position on both strands, as opposed to enzymes that make staggered cuts.

blunt-end ligation   A cloning technique in which both the vector (see CLONING VECTOR) and insert to be spliced into the vector have blunt ends that must be joined by the enzyme ligase. Such ligation is more difficult to achieve than one in which the vector and insert have complementary single-stranded overhangs, which first form hydrgen bonds before the ligation step.

B-lymphocytes   The antibody-producing cell of the humoral immune response. When stimulated with antibody, these cells divide and differentiate into plasma cells that secrete antibodies.

branch migration   A proposed step in the process of DNA recombination, or DNA crossing over, in which there is movement of the crossover point of the recombinant intermediate.

breakage and reunion   Physical breakage of DNA molecules and rejoining of parts of two different molecules, resulting in recombination or crossing over.

5-bromouracil (5-BU)   A chemical that causes mutations in DNA because it resembles thymine, a natural constituent of DNA. When incorporated into DNA in place of thymine it can readily pair with guanine. In its presence, an A-T base pair is replaced by a G-C base pair after two rounds of replication. This is called a transition mutation.

broth   A liquid culture medium for microorganisms.

buffer   A substance in a liquid that tends to resist changes in pH by absorbing hydrogen ions or hydroxyl ions.

Burkitt's lymphoma   A relatively common tumor in East Africa and New Guinea, but rare in other parts of the world. The Epstein–Barr virus (EBV), the etiological agent of infectious mononucleosis, is associated with this disease, but it is not currently known whether the relationship of the virus to the disease is casual or causal.

bursa of Fabricius   A lymphoid organ of the chicken responsible for the maturation of B lymphocytes. The B cells were so named because of this organ. However, humans and other mammals do not possess a bursa, and its equivalent in these organisms is probably other lymphoid tissues, such as the tonsils, appendix, Peyer's patches, and the lymphoid follicles.

burst number   The number of viral particles produced per cell after infection.

C

C600   A strain of *E. coli* commonly used in genetic experiments and as a host for cloned PLASMIDS.

CAAT box   A consensus nucleotide sequence (see CONSENSUS SEQUENCE) that has homology to GGT (orC) AATCT, is found in the promoter region of many eukaryotic genes, and is required for efficient transcription.

calmodulin   A ubiquitous calcium-binding protein that serves as an intracellular receptor of Ca^{2+} and, in its active form, mediates an intracellular response to Ca^{2+} as a second messenger.

calorie A unit of energy measurement: the amount of energy needed to raise the temperature of 1 g of water 1°C.

cancer A class of diseases in which normal cell control is lost so that certain cells in the body proliferate uncontrollably, invade other tissues, and spread to distant sites (metastases) in the body.

capped 5'-ends A methylated guanosine residue attached to the 5'-end of a eukaryotic mRNA. The bond is made between the 5'-phosphate group of the nucleotide and the 5'-end of the RNA, so the nucleotide is called inverted. This cap may give stability to the mRNA.

capping of mRNA The posttranscriptional process that adds a guanosine residue to the 5'-end of a eukaryotic mRNA and then methylates it.

capsid The protein coat of a virus.

capsomere The protein subunits of the capsid.

capsule An envelope of CARBOHYDRATE or a slime layer surrounding some microorganisms. Capsules contribute to the invasiveness of some bacteria because they enable the organisms to evade PHAGOCYTOSIS.

carbohydrate A sugar or the name for molecules that contain carbon, hydrogen, and oxygen in the ratio $C_nH_{2n}O_n$ and that can be simple monomers, such as glucose or fructose, disaccharides, or two molecules joined by a glycosidic bond (see GLYCOSIDIC LINKAGE), such as sucrose (common table sugar) or lactose (milk sugar), or polymers, called polysaccharides, containing up to thousands of simple sugar molecules, such as starch, cellulose, and glycogen.

carbon dioxide cycle The flow of CO_2 from organisms that can photosyn-

thesize (plants and algae) and convert CO_2 into organic foodstuffs to all other organisms that consume the organic molecules and give off CO_2 as waste product.

carbon fixation The process by which plants and algae (photosynthesizers) convert inorganic carbon, CO_2, into organic molecules, specifically carbohydrates, which are used as food for other organisms.

carbon source Any organic carbon-containing molecule that can be metabolized to produce energy in the form of ATP in an organism. In general, carbohydrates serve as carbon sources for most organisms, but some amnio acids can also be utilized for energy production.

carbonyl group The atoms of carbon and oxygen, in which the oxygen is bonded to the carbon via two chemical bonds $C=O$.

carboxyl group The atoms of carbon, oxygen, and hydrogen, in which one oxygen is bonded to the carbon via a double bond, and the oxygen and hydrogen form a hydroxyl group (OH) and are bonded to the carbon via a single bond with the oxygen.

$$-C=O$$
$$|$$
$$OH$$

Carboxyl groups are found on organic acids.

carboxyl terminus The end of a molecule where a carboxyl group is found. Proteins made up of amino acids have a carboxyl terminus and an amino terminus.

carboxypeptidases Enzymes that remove successive amino acids from proteins, starting at the carboxy terminal end (see CARBOXYL TERMINUS), by hydrolyzing the peptide bond between amino acids.

carcinoge A   cancer-producing agent, or any chemical or physical agent that can produce a tumor in an animal or cause normal cells in culture to become transformed.

carcinogenesis The process by which a normal cell transforms to a cancerous one. (See TRANSFORMATION.)

carcinoma A tumor derived from epithelial cells.

cardiac muscle The muscle tissue of the heart responsible for pumping blood through the body's circulatory system. It is striated and looks very similar to skeletal muscle, but uses different carbon sources for energy production.

β-carotene A carotenoid pigment that harvests light energy and transfers this energy to other photosensitive pigments, such as chlorophylls, in a photosystem. This pigment gives a red or orange color to carrots, tomatoes and other plants.

carotenoids A family of pigments that can absorb a range of wavelengths of light and funnel the energy to chlroplyll a, the major photosynthetic pigment of plants. Thus, these accessory pigments greatly extend the range of light wavelengths that can be used for photosynthesis.

carrier A substance involved with the transport of materials.

carrier protein A protein, embedded within the cell membrane, that binds to a specific compound or group of related compounds and aids in transporting it from the outside of the cell through the membrane lipid bilayer to the interior of the cell.

casamino acids A mixture of amino acids that results from the enzymatic breakdown of the milk protein casein.

cascade A series of reactions triggered by one reaction or compound.

casein hydrolysate The breakdown product of the milk protein casein to its constituent amino acids by either enzymes or acid hydrolyzing the peptide bonds between the amino acids.

catabolic pathway A series of reactions that break down compounds to simpler ones, usually with the release of energy trapped into high-energy molecules such as ATP.

catabolism The degradation of complex substances to simple ones.

catabolite A substance that can be broken down by an organism to yield energy, usually in the form of ATP.

catabolite activator protein (CAP) or **catabolite repressor protein (CRP)** A protein that, when bound to cAMP, will bind to the promoter region of some operons encoding enzymes that metabolize sugars in bacteria to enhance transcription. Thus, the protein bound to cAMP acts as a positive regulator of transcription. (See LAC OPERON.)

catabolite repression A process found in certain bacteria in which there is decreased synthesis of enzymes involved in catabolism when a preferred alternative catabolite is present. For example, the enzymes that metabolize the sugar lactose are not synthesized by bacteria when they are grown on the sugar glucose.

catalase An enzyme that breaks down hydrogen peroxide, a toxic waste product of metabolism, to water and oxygen. It is usually found in the microbody of PEROXISOME organelles.

catalysis (catalytic) The acceleration of a reaction by a catalyst.

catalyst A substance or physical agent that speeds up a reaction but is not consumed during the course of the reaction. Catalysts in biochemical reactions are enzymes. Catalysts change the rate at which reactions approach equilibrium, but do not affect the position of equilibrium.

catalytic site The location of an enzyme where the active site is, or the place where substrates bind and the reaction proceeds. The catalytic site brings the reactants close together, eliminating the need for random collisions and thus making the reaction more efficient.

catenation Linking multiple copies of a macromolecule.

cation A positively charged ion.

C banding A technique for staining the highly repeated DNA sequences in the region of the chromosome surrounding the CENTROMERE. (See SATELLITE DNA.)

C$_3$ cycle The Calvin cycle or that part of photosynthesis where CO_2 is fixed to form a three-carbon organic compound that is subsequently converted to a six-carbon sugar.

C$_4$ cycle The Hatch–Slack pathway, or an accessory very efficient pathway to fix CO_2 used by plants that grow in hot dry climates with low CO_2 levels.

cell The smallest membrane-bound unit capable of replication. Cells may function independently, such as those of unicellular microorganisms, or cooperatively, such as tissue or organ cells.

cell coat A layer of carbohydrates that protrudes into the extracellular space from the cell membrane in animal cells and is seen in the electron microscope as an electron-dense coat on the surface of the cell.

cell culture A population of cells grown in a medium.

cell cycle A sequence of events involved in the replication of the genetic material of the cell and the orderly parceling of it to two daughter cells. The cell cycle consists of the G1, S, and G2 phases, which make up the interphase, and the M phase, or MITOSIS, where chromosome division occurs.

The Cell Cycle

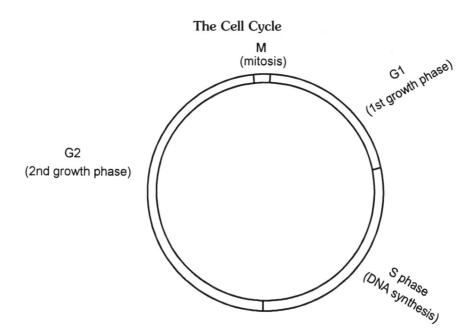

cell-division-cycle (cdc) genes Any of approximately 50 genes that control the cell cycle in yeast.

cell-division-cycle (cdc) mutant Temperature-sensitive cdc mutants of yeast that become blocked or show aberrant behavior in various parts of the CELL CYCLE at a temperature at which the mutation can be expressed (the restrictive temperature).

cell fractionation The process of preparing a cell-free extract and dividing the cell contents into fractions by centrifugation techniques.

cell-free extract The product of treating a suspension of cells with a substance(s) that destroys the cell wall (in bacteria and plants) and/or the cell membrane, thus releasing the cytoplasm and cell organelles. Sometimes the cell-free extract refers to the soluble portion of the internal cellular contents after removal of the organelles and cell membrane debris.

cell-free protein synthesis The synthesis of proteins in a test tube using a cell-free extract to supply the necessary enzymes and components and depending on addition of amino acids and mRNA.

cell fusion The process of fusing two different cells, first creating a heterokaryon that contains both of the nuclei, and then a fusion of the nuclei to create a synkaryon. The fusion occurs by reaction between the two cell membranes, which is brought about by treatment with SENDAI VIRUS or polyethylene glycol.

cell line A cell culture started from a particular type that can be cultured indefinitely in the laboratory, and thus characterized as "immortal."

cell lineage A complete set of ancestral cells and cell divisions that makes up a certain cell type during development.

cell-mediated immune system See CELL-MEDIATED IMMUNITY.

cell-mediated immunity A type of immunolgical response mediated by cytotoxic T lymphocytes, or killer T cells, and is used by the body to destroy cells carrying foreign antigens, such as virally infected cells, tumor cells, and non-matching-tissue grafts.

cell membrane The boundary that separates the cell contents from its environment. It is composed of a phospholipid bilayer associated with proteins embedded in the bilayer (intrinsic or transmembrane) or external to it (extrinsic). The cell membrane provides a selectively permeable barrier to the cell, allowing substances through that are needed by the cell and preventing leakage of important substances.

cell plate The boundary between two newly formed nuclei in a plant cell that is about to divide into two daughter cells. It consists of cell-wall material and a cell membrane that grows and eventually becomes contiguous with the existing cell wall and cell membrane. Also called the phragmoplast.

cell sorter An instrument used to separate and analyze different classes of cells from mixed populations. The fluorescence-activated cell sorter (FACS) separates different cell types in a population based on external antigens that bind to antibodies labeled with fluorescent dyes.

cell sorting The process of sorting different cell types in a heterogeneous population. (See FLUORESCENCE-ACTIVATED CELL SORTING.)

cell synchronization The process by which all cells in a population come to be in the same phase of growth and consequently undergo cell division simultaneously.

cell theory The theory that states that the cell is the basic structural unit of all organisms and that all cells arise from preexisting cells.

cellular oncogene See C ONCOGENE.

cellulase An enzyme that hydrolyzes cellulose (a polymer consisting of glucose units) to cellobiose, a disaccharide consisting of two glucose units.

cell wall The rigid or semirigid layer peripheral to the cell membrane of bacteria, algae, fungi, and plants. In plants the cell wall is composed of microfibrils of cellulose embedded in a matrix. The bacterial cell wall, the peptidoglycan layer, is a complex structure of chains of alternating residues of N-acetylmuramic acid and N-acetylglucosamine held together by peptide bridges.

centimorgan One hundredth of a morgan, the unit of genetic distance or a map unit distance, named in honor of Thomas Morgan's contribution to mapping genes in *Drosophila.*

central dogma The concept that all genetic information flows from DNA. Information in DNA is passed on to progeny cells by DNA replication, and information stored in DNA can be transcribed to RNA, which is then translated to synthesize proteins.

central nervous system The sensory and motor cells (neurons) of the brain and spinal cord.

centrifugal force The force that tends to impel substances outward from a center of rotation.

centrifuge An instrument that separates substances from liquids by centrifugal force and separates substances from other substances based on how each moves in a centrifugal field.

centriole A structure, composed of microtubules, found in the nucleus and involved in the formation of the spindle apparatus, which aids in the orderly parceling out of duplicated chromosomes to daughter cells during cell division. The centriole looks exactly like the basal body, the organelle embedded at the periphery of the cell and that serves as the base of the cell's locomotive appendages, the flagella and cilia.

centromere The point along the chromosome to which duplicated sister chromatids are joined before the chromosomes are divided into two daughter cells. It also serves as the site of attachment of the kinetochore, the structure on which microtubules of the spindle apparatus attach to pull the duplicated chromosomes to opposite ends of the cell during MITOSIS.

centromeric sequences Special nucleotide sequences in the DNA of chromosomes that serve as sites where the spindle apparatus attaches to chromosomes during mitosis. (See YEAST ARTIFICIAL CHROMOSOMES.)

centrosome The cell center or a microtubule organizing center consisting of granular material surrounding two CENTRIOLES.

cephalosporin C One of a group of antibiotics, the cephalosporins, produced by the fungus *Cephalosporium,* and which resembles penicillin in structure and mode of action.

cerebellum Part of the hindbrain consisting of two hemispheres and a small central portion.

cerebrospinal fluid (CSF) Fluid produced in brain ventricles, which fills the ventricles and the central canal of the spinal cord, to cushion the brain and protect it from blows to the skull or bruises resulting from sudden movements of the head.

cerebrum That part of the brain located above and in front of the hindbrain, consisting of a pair of hollow, convoluted lobes.

cesium chloride (CsCl) gradient centrifugation. A method used to separate and/or purify molecules, usually nucleic acids. The nucleic acids to be separated are mixed with an appropriate amount of cesium chloride, a dense, chemically inert salt, and centrifuged at high speeds for hours to days. The cesium chloride establishes a density gradient during the centrifugation, and the molecules of nucleic acid move up or down the gradient to reach their position of buoyant density in the gradient.

channel protein   A cell membrane-embedded protein, part of a channel structure allowing substances of appropriate size and charge to pass through the membrane by diffusion.

Charon phage   A vector, used for cloning DNA, constructed from bacteriophage λ. *Charon* is named from the ferryman Charon, who, in the ancient Greek myth, transported the spirits of the dead across the River Styx.

CH$_3$ choline   A small alcohol of the structure $HO-CH_2-CH_2-N(CH_3)_2$ found in membrane CH$_3$ phospholipids and part of the important neurotransmitter acetylcholine.

chelator   An organic compound in which atoms form bonds with metals, thus removing free metal ions from solution.

chemiosmotic theory   A model, proposed by Peter Mitchell, that couples electron transport to oxidative phosphorylation (ADENOSINE TRIPHOSPHATE synthesis during respiration) or photophosphorylation (ATP synthesis during photosynthesis). It postulates that the energy needed to drive the synthesis of ATP is stored in a proton gradient across the inner membrane of the MITOCHONDRION or the thylakoid membrane of the CHLOROPLAST, and this gradient forms during electron transport. When the gradient is relieved by the transport of protons across the membrane, the stored energy is used to drive the synthesis of ATP.

chemoautotroph   An organism that obtains its energy from the oxidation of chemical bonds, usually the oxidation of inorganic metal ions, and can use inorganic carbon, or CO$_2$, to make biological molecules.

chemolithotroph   A synonym for chemoautotroph.

chemoorganotroph   An organism that obtains its energy from the oxidation of chemical bonds and requires organic carbon compounds for growth. A heterotroph.

chemostat   An apparatus used to maintain a bacterial culture in continuous culture or exponential growth, by coordinating the rate of addition of some limiting nutrient to the rate of removal of spent medium and cells.

chemotaxis   The movement of an organism to an attractant and away from a repellant.

chemotherapy   The treatment of a disease with chemicals, but the term is usually used to define the treatment of cancer with drugs that selectively kill faster-growing tumor cells.

chemotroph   An organism that obtains its energy from the oxidation of chemical bonds. (See CHEMOAUTOTROPH, CHEMOORGANOTROPH.)

chiasma (chiasmata, pl.)   The location of a crossover event between two chromatids in the tetrad structure of synapsed duplicated pairs of chromosomes, which occurs during prophase I of MEIOSIS.

chimera   An animal formed from aggregates of genetically different groups of cells. Chimeras are made by combining early-stage embryos that arise from fertilized eggs of two different sets of parents, or by injecting cells from an early embryo of one GENOTYPE into the blastocyst of another genotype. The term comes from mythology, where the Chimera was a creature with the head of a lion, the body of a goat, and the tail of a serpent. (See BLASTULA.)

chimeric DNA   A recombinant DNA molecule, or one in which a fragment of DNA from one source is spliced into a vector from another source.

Chi Structure

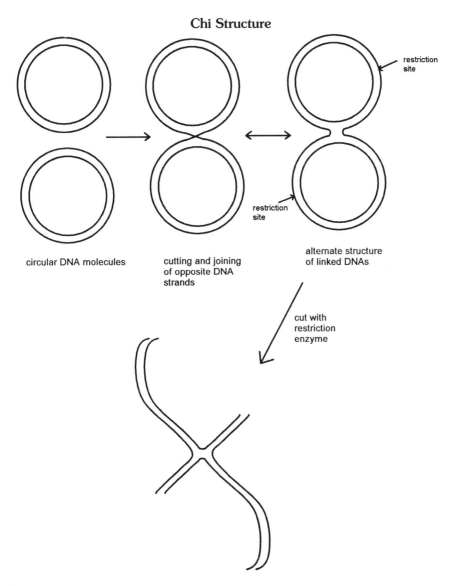

circular DNA molecules

cutting and joining
of opposite DNA
strands

alternate structure
of linked DNAs

restriction
site

restriction
site

cut with
restriction
enzyme

chiral compound A compound, usually a carbon compound, that is optically active—one that has the ability to rotate the plane of polarized light either to the left or to the right, due to its ability to exist in one of two mirror images. (See ENANTIOMERS.)

chirality (handedness) The nonidentity of a chemical compound with its mirror image. (See ENANTIOMERS.)

chi sequence A sequence of bases on the genome of the bacterium *E. coli* that signals a nuclease to cut at that site for recombination or crossing over to occur. It serves as a hotspot of recombination because it is used preferentially as a site where recombination occurs.

chi structure The structure generated when the figure-eight-shaped molecule, which is an intermediate form in the process of recombination between two circular DNA molecules, is cut by a restriction enzyme that cuts each circular DNA once. It is so named because the four-armed structure, as seen by electron microscopy, resembles the Greek letter X.

chitin   The structural POLYSACCHA-RIDE present in the EXOSKELETON of insects, in the cell walls of fungi, and in crustacean cells composed of units of *N*-acetylglucosamine.

chloramphenicol   An antibiotic, produced by *Streptomyces venezuela,* that inhibits protein synthesis in bacteria, mitochondria, and chloroplasts, but not in higher organisms. It is used to amplify recombinant DNA molecules when a PLASMID VECTOR is used to make the recombinant. Chloramphenicol specifically inhibits the host cell's replication because it depends on new protein synthesis, but will not inhibit certain plasmid replication. Thus, in the presence of the antibiotic, the recombinant molecule preferentially replicates up to 200 copies per cell.

chloramphenicol acetyl transferase (CAT) assay   An assay for determining whether a given DNA fragment may contain PROMOTER activity by ligating the fragment to the CAT gene in an expression vector and observing whether the CAT enzyme is made when the vector is transfected into animal cells. (See AMINOGLYCOSIDE-3-PHOSPHOTRANS-FERASE, CLONING VECTOR, LIGATION.)

chloramphenicol acetyl transferase (CAT) gene   A bacterial gene that catalyzes the transfer of an acetyl group to chloramphenicol. The CAT gene is commonly used as a reporter gene in experiments designed to demonstrate that certain DNA sequences can function as promoters.

chlorophyll   A light-absorbing pigment found in the chloroplasts of plants and algae that is essential as an electron donor in the process of photosynthesis. It gives the green color to plants.

chloroplast   The organelle in plant cells and algae responsible for photosynthesis. It contains the chlorophyll and the proteins used to carry out the reactions of photosynthesis.

chloroplast DNA (ctDNA)   DNA found in the chloroplast.

cholesterol   A major lipid constituent of cell membranes of higher organisms and a precursor in steroid synthesis.

cholinergic neuron   Pertaining to the general class of neurons that utilizes acetylcholine as a NEUROTRANSMITTER.

chondroitin sulfate   A sugar acid that is frequently a component of the fuzzy layer, an extracellular layer of collagen and glycosaminoglycans peripheral to the CELL COAT of some animal cells.

chromatid   One of the daughter duplicated chromosomes, joined at the CENTROMERE to its sister chromatid and seen at the prophase and metaphases stages of MITOSIS and MEIOSIS.

chromatin   The material of the chromosome, consisting of DNA associated with histone proteins.

chromatographic techniques   A group of techniques for separating substances in a mixture based on the way the component substances bind to, or dissolve in, various solids, liquids or gases. In a chromatographic separation the mixture is usually present in a liquid or gas phase that moves to a second, usually solid or liquid phase in which the individual components separate from one another. The term chromatography is derived from the Greek word *chroma* (color) and stems from the original use of the technique for separating mixtures of colored substances.

column chromatography   The separation or purification of substances, generally proteins, based on their specific binding to a column prepared by attaching a ligand that will specifically bind the substance to some solid support and washing the support-filled column with a liquid that will compete with the bound substance for the support material in the column.

gel filtration chromatography The separation of molecules, generally proteins, based on their sizes, as seen by their flow through a column prepared with porous beads of a carbohydrate polymer that traps smaller molecules, impeding their flow, but permits larger molecules to flow rapidly. Also called molecular sieving.

high-performance liquid chromatography (HPLC) A chromatographic method in which a high resolution of separation is achieved by improvements in the packing of columns and the flow of solvents through the columns under high pressure. The method yields very sharp peaks of substances eluted from the column.

hydroxyapatite chromatography The separation of molecules based on their relative binding to a column prepared with calcium phosphate. Such a column is used to separate double-stranded DNA, which will bind to it, from single-stranded DNA, which will pass through it, as well as proteins.

ion-exchange chromatography The separation of substances based on their charge and, thus, their affinity to a column prepared with a charged support material. Substances are eluted from the column with a solution of ions that compete with the substances binding to the column.

paper chromatography The separation of substances based on their relative solubilities in a mixture of solvents. Substances to be separated are applied to a paper support, and the solvents travel up or down the paper via capillary action, dissolving and transporting the substances on the paper.

reverse-phase chromatography The separation of substances based on their relative hydrophobicities. The support matrix is prepared so as to contain large hydrophobic carbon chains that will bind hydrophobic proteins more strongly, and these proteins are thus eluted from the column more slowly than hydrophilic ones.

thin-layer chromatography The same as paper chromatography, but the support is a glass plate coated with a silica gel.

chromatography An analytical technique used to separate molecules from each other based on differences in their affinities for and/or migration on some support when subjected to the flow of a solvent. Chromatographic methods differ with respect to the nature of the solid support and the type of mobile phase (solvent). (See CHROMATOGRAPHIC TECHNIQUES.)

chromogenic label Any molecule, attached to a biological probe molecule, that generates a colored compound(s) as a means of visualizing the location and amount of probe bound to a particular target.

chromomeres The beadlike structures on LAMPBRUSH CHROMOSOMES seen during meiosis when the chromosomes become extended.

chromosomal mutation A change in the sequence of base pairs of the DNA encoding a gene, resulting in a change on the protein encoded by that gene. The change can be as simple as a change in one base (missense mutation involving one amino acid on the protein) or the addition or deletion of one base (resulting in a change in the reading frame, thus affecting many amino acids on the protein) to more complex changes, such as the addition or deletion of many bases or the transposition of part of one chromosome to the other.

chromosome The structure in the nucleus containing the genetic information and composed of DNA and HISTONE proteins associated with DNA. Also refers to the gene-containing unit of bacteria, viruses, mitochondria, and chloroplasts, although these do not resemble the chromosomes of higher organisms in structure or histone content.

chromosome mapping The techniques used to assign specific gene locations on the chromosome based on crossover frequencies and linkage frequencies between genes.

chromosome puffs The uncoiled regions of DNA found on the giant POLYTENE CHROMOSOMES of the salivary glands of certain members of the *Diptera* group (e.g., fruit fly) with the appearance of puffs when observed by conventional light microscopy. These regions have been shown to be sites where the chromosomal DNA is actively in the process of transcription.

chromosome walking, jumping, crawling A procedure for locating a gene by using cloned genes close to the target, preparing PROBES from these genes, and using them to isolate members of a genomic LIBRARY that hybridize to the probe but contain other genetically linked material. If each member of the library contains an insert of 10,000 base pairs (bp), a probe that can hybridize to the first 100 bp can be used to isolate a gene located 9000 bp away. If the gene is located more than 10,000 bp away, then the first probe is used to isolate a clone to make a second probe, which can be used to isolate a third probe until the specific gene is found. This procedure is called chromosome walking, because probes are isolated and then used to identify portions of the chromosome that are contiguous to each other.

chymotrypsin A digestive enzyme, found in the small intestine, that hydrolyzes peptide bonds, thus cleaving proteins into their component amino acids.

cilium (cilia, pl.) Short, hairlike, membrane-bounded appendage composed of microtubules used in the locomotion of cells.

circadian clock A biological timing mechanism that controls a type of natural synchrony (see CELL SYNCHRONIZATION) by controlling cell division.

cis A term used in genetics to define an event or gene whose action occurs on the same chromosome.

cis acting Pertaining to a genetic element exerting an effect on a target located within the same unit. For example, a promoter element is said to be cis acting with respect to the genes it controls when both are on the same strand.

cis-acting gene A regulatory gene that controls transcription of genes that lie near it on the same chromosome by binding protein factors needed to turn transcription on or off. (See CIS-TRANS TEST.)

cis face The portion of the Golgi stack of vesicles that has just formed (also called the forming face) and is oriented toward the rough ENDOPLASMIC RETICULUM. (See GOLGI APPARATUS.)

cisterna A flattened membrane-bound sac, such as is found in the ENDOPLASMIC RETICULUM.

cis–trans test A test historically used to determine whether two mutations both map in the same functional genetic unit (i.e., a gene; see CISTRON) by genetic complementation. The test is performed by mating two organisms, each carrying a different mutation and each of which is known to affect a testable function (e.g., the ability to make some nutrient). If the offspring of the cross are able to carry out the function, the two mutations are said to lie within separate functional units, whereas if offspring are unable to carry out the function the two mutations must lie within the same functional unit. In higher organisms the "functional unit" may be an entire chromosome rather than a gene. If two mutations are found on separate chromosomes, they are in the trans configuration; if they are on the same chromosome, they are in the cis configuration. Complementation only occurs between trans mutations in different genes, not on the same gene.

cistron In prokaryotic organisms, a genetic unit, later called a "gene," defined in terms of clusters of mutations that all lie within a common functional unit as determined by the cis–trans test.

citric acid An organic acid containing three carboxyl groups and an important intermediate in a cyclic pathway called the Krebs cycle, tricarboxylic acid (TCA) cycle, or citric acid cycle, and which is responsible for the metabolism of glucose to water and carbon dioxide in the presence of oxygen.

clathrin A large protein that forms a basketlike structure around vesicles that transport molecules into or through cells, or at sites (coated pits) where ENDOCYTOSIS will occur.

cleavage (1) The breaking of bonds between units of macromolecules, such as the enzymatic cleavage of amino acids from protein. (2) The furrowing that occurs in animal cells to form two daughter cells from a parent cell after mitosis when the chromosomes have divided. (3) A series of cell divisions that occur during early animal embryogenesis.

clinical trials Testing of new drugs or therapies on humans in a rigorous controlled setting.

clonal deletion The selective loss, early in development, of B and T cells of the immune system that produce antibodies or have receptors for antigens that are an integral part of the organism (self-antigens). This process is necessary to prevent the immune system from attacking the cells and tissues of the organism later in life (autoimmunity).

clone bank (clone library) A collection of recombinant DNA molecules of the genomic material of a particular organism, prepared by fragmenting the DNA of the organism and splicing each of the fragments into vector molecules. Also known as a library.

clone (cellular) A population of cells that have been derived from the divisions of one cell, so the population is genetically identical.

clone (DNA) Recombinant DNA molecule or recombinant molecule. A gene or fragment of DNA that has been spliced into a VECTOR so that the DNA can be amplified many times by transferring the recombinant molecule into a host organism (usually a bacterium or yeast) that can be grown in large quantities.

cloning The process of creating, isolating, and amplifying a recombinant DNA molecule. (See CLONE, CLONE BANK, CLONING VECTOR.)

cloning vector The molecule of DNA used to house the DNA fragment to be cloned. Vectors are small chromosomes, either PLASMID or BACTERIOPHAGE, capable of self-replication in a host cell and producing many copies of itself per host cell, thus amplifying the number of copies of the cloned fragment.

Clostridium The genus of organisms that are obligate anaerobes and produce spores. Members of this group produce powerful toxins and are responsible for diseases such as botulism, gas gangrene, and tetanus.

coated pit An invaginated site on the cell membrane lined with clathrin facing the interior of the cell and containing specific receptors at the exterior where molecules interact with the receptors for transport into the cell via receptor-mediated ENDOCYTOSIS.

coated vesicle   Small, membrane-bound droplets, coated with a basket of clathrin, transporting molecules from the outside of the cell via receptor-mediated endocytosis, having arisen from coated pits, or transporting newly made proteins to be either sorted to organelles, or secreted to the outside of the cell.

coat protein(s)   Proteins that make up the outer layer, or coat, of a virus.

coccus (cocci, pl.)   The name for a type of bacterial cell with a round morphology.

code   Refers to the way the genetic information is stored in the DNA. (See GENETIC CODE.)

coding strand   The strand of DNA used as a template to make mRNA. It contains the complement of the code to be translated.

codon   The sequence of three consecutive nucleotide bases that specifies or ''codes for'' a particular amino acid. (See DEGENERATE CODE.)

coenzyme (cofactor)   A small non-protein organic molecule associated with an enzyme (apoenzyme) and required for catalytic activity. The coenzyme plus the apoenzyme is called a holoenzyme. Although the apoenzyme does not change during the course of catalysis, the coenzyme may be chemically altered, but it is regenerated and reused in subsequent reactions. A number of vitamins serve as components of coenzymes. For example, two common coenzymes involved in energy metabolism are nicotinamide adenine dinucleotide (NAD^+) and flavin adenine dinucleotide (FAD). Both the nicotinamide and flavin portions of the molecules are derivatives of the B vitamins nicotinic acid and riboflavin.

coenzyme A (CoA or CoASH)   A small organic molecule composed of adenosine diphosphate linked to the vitamin pantheteine phosphate, which serves as a carrier of acyl groups. CoA is particularly important as an acyl carrier during the oxidation of sugars for energy production.

cofactor   A metal ion, such as Mg^{2+}, Fe^{3+}, or Mn^+, or COENZYME that functions in association with enzyme proteins and that are necessary for complete enzymatic activity.

Cohen, Stanley (b.1922)   Molecular biologist who carried out the first cloning experiments by splicing the gene encoding resistance to the antibiotic tetracyclin from one strain of bacteria *(Staphylococcus aureus)* into a plasmid from another strain *(Escherichia coli)* in a test tube. The recombinant molecules were transferred into cells of *E. coli,* and transformed cells with tetracycline resistance grew into colonies. These experiments demonstrated that genes isolated from one organism, spliced into a vector, and transferred into a host organism are intact and capable of producing functional proteins.

cohesive ends (sticky ends)   The single-stranded extensions of a double-stranded DNA molecule that show complementarity to other single-stranded extensions of DNA molecules. Such sticky ends are generated by restriction endonucleases.

coincidental evolution (concerted evolution)   In genes that have become duplicated, the tendency for mutations occurring in one copy to appear in the other, with the result that the effects of evolution appear in both copies at the same time.

cointegrate structure   A molecule of DNA in which a TRANSPOSON has mediated the joining of two PLASMIDs, with copies of the transposon occurring at the joints between the two plasmids.

This is the first step in the transposition of the transposon from one plasmid to another.

colcemid A drug that blocks microtubule formation and thus disrupts events, such as chromosome separation during mitosis, depending on microtubule function.

colchicine A drug that disrupts microtubule function as does colcemid.

Col E1 A naturally occurring plasmid that is carried by some strains of *E. coli,* and has been used as a basis for constructing a number of cloning vectors for making recombinant DNA molecules. It is one of a family of plasmids, which encodes genes for a colicin, and immunity proteins, which protect Col-harboring cells from the bacteriocidal effects of the colicin it produces.

colicin An antibiotic encoded by certain *E. coli* plasmids, such as Col E1. Colicins kill bacteria by various mechanisms, including inhibition of protein synthesis, inhibition of active transport, and DNA degradation.

coliforms A group of bacteria including *Escherichia, Kelbsiella, Enterobacter, and Citrobacter,* which are small, rod-shaped, facultative anaerobes, stain gram-negative, and ferment lactose with gas production within 48 hours of growth. They are used to assess fecal pollution of water.

colinear Having the linear array or sequence of one molecule correspond to that of another. The sequence of bases found on the mRNA (see MESSENGER RNA) corresponds to the sequence of amino acids found on the protein that it encodes. This relationship extends to the sequence of bases on the DNA for bacteria, but in higher organisms the DNA also contains some intervening sequences (see INTRONS) that must be eliminated before obtaining colinearity.

coliphage A bacterial virus that infects and reproduces in coliforms.

collagen A fibrous protein that is a major component of connective tissue, and is found in the fuzzy layer that envelopes animal cells.

colloid A suspension of microscopic particles ranging in size from 1 nm to 1 μm, dispersed in some medium. Hydrophylic colloids are composed of macromolecules that remain dispersed in an aqueous solution, because of the affinity of the particles for water. Hydrophobic colloids are less stable and are composed of insoluble particles suspended in water and remaining in a suspended state due to repulsive forces among particles.

colony A group of cells growing from a single cell on some solid medium, such as an agar plate.

colony counter An instrument used to count the number of colonies on an agar plate. Manual counters have an electronic stylus that creates a signal that is counted when touched to a colony. Automatic colony counters have scanners that detect density differences and can read an entire plate for the total number of colonies.

colony-forming unit (CFU) A viable cell that gives rise to a colony.

colony hybridization A method used to identify colonies harboring a particular gene or DNA sequence. Colonies on an agar plate are partially transferred to a membrane, generally nitrocellulose or nylon, by gently pressing the membrane on top of the colonies. The membrane is treated with alkalai to denature the DNA in the cells, heated to fix the DNA on to it, then washed with a labeled PROBE to identify those colonies that carry the sequence. Once the colony is identified on the membrane, it can be picked from the original plate and cultured to study or to further isolate the gene or sequence.

colorimeter An instrument that quantitates the amount of substance in solution by measuring the amount of light, at a given wavelength, absorbed by the solution. Colorimetry is based on the BEER–LAMBERT LAW defining the extinction coefficient of a substance and the relationship of light absorbed by a substance to its concentration. Colorimeters are also used to measure the turbidity of solutions, which is an indication of the number of particles in suspension or the number of bacterial cells in a culture.

combining site The site on an antibody molecule where the antigen interacts.

commensalism A relationship between members of different species living within the same cultural environment with one organism benefitting from the relationship and the other unaffected.

compatibility group Defining a group of plasmids on their ability to co-exist in the same cell with another PLASMID from a different group.

competence The state of a bacterial cell that has the ability to take up DNA from the environment. Some species of bacteria develop natural competence by synthesizing competence factors and DNA receptor proteins, which aid in the uptake of DNA into the cell. Other species, such as *E. coli,* can be made competent by treatment of cells with high concentrations of $CaCl_2$ in the cold.

competition hybridization A technique for determining the degree of similarity between two nucleic acids by measuring the degree to which the two nucleic acids hybrize to one another in the presence of a third nucleic acid, which acts as a standard.

competitive inhibition The inhibition of an enzyme by a substance that reversibly binds to the active site of the enzyme and thus competes with the substrate for the site.

complement A group of serum proteins activated by reaction with antigen-antibody complexes. Once activated, they aid in the killing of pathogenic bacteria and/or facilitate PHAGOCYTOSIS.

complementary base pairing The formation of hydrogen bonds between adenine and thymine and between guanine and cytosine. (See COMPLEMENTATION.)

complementary base sequence A sequence of bases that can form hydrogen bonds with another sequence. (See COMPLEMENTARY BASE PAIRING.)

complementary DNA (cDNA) The single-stranded, complementary DNA copied from messenger RNA (mRNA) by the enzyme reverse transcriptase.

complementary DNA (cDNA) cloning A recombinant DNA technique in which double-stranded cDNA is spliced into a vector so the gene can be amplified or expressed.

complementary DNA (cDNA) library A collection of cDNA molecules spliced into vectors, made by using all the mRNA molecules in a cell or organism and copying it with REVERSE TRANSCRIPTASE. The LIBRARY is subsequently screened with appropriate probes to pick out the clone of choice. (See CLONE LIBRARY, CLONING VECTOR.)

complementation (1) The ability of one chain of polynucleotides (either DNA or RNA) to form hydrogen bonds with another chain, because of a coincidence of adenine-thymine pairs of bases and guanine-cytosine pairs of bases on each strand. (2) Genetic complementation is the ability of one mutant to supply a required function to another mutant (see CIS-TRANS TEST). (3) Cloning by complementation is a technique in which a mutant host cell (e.g., which lacks the ability to synthesize some nutrient) is infected with a library and a clone is picked that has the ability to synthesize the nutrient. This clone is derived from a mutant cell that picked up a recombinant molecule containing a functional gene having the ability to replace its own faulty one.

complementation test See CIS–TRANS TEST.

complement–fixation (CF) test A serological test for antibodies based on the ability of complement to LYSE red blood cells. Serum to be tested is mixed with antigen and complement. An indicator system of sheep red blood cells (RBCs) and antibody against the sheep RBCs is added. If specific antibody for the antigen is present in the serum, it will combine with the antigen, bind complement, and no complement will be available to lyse the sheep RBCs. Thus, no lysis of sheep RBCs indicates the presence of antibody in the serum in a complement-fixation test.

complete medium A culture medium that supplies all the nutrients (amino acids, vitamins, and bases found in nucleic acids) an organism needs for growth.

complexity A measure of the number of different base-pair sequences on a given genome.

composite transposon A transposable element (see TRANSPOSON) made up of insertion sequence (IS) elements flanking a central portion of DNA sequence that usually contains a gene or genes encoding antibiotic resistance determinants. Common composite transposons are Tn5 and Tn10.

compost A mixture of decaying organic material used for fertilization or rejuvenation of soil.

concanavalin A (con A) A lectin, isolated from the jack bean *(Canavalia ensiformis),* that binds to certain sugar residues. It is used in AFFINITY CHROMATOGRAPHY to purify glycoproteins, and is also used to agglutinate cells by cross-linking glycoproteins found at the cell surfaces. In addition, con A induces resting lymphocytes to divide.

concatamer A series of the same DNA molecules linked in tandem, thus creating a dimer, trimer, or multimer.

C-oncogene A normal cellular gene that has a viral oncogene, or tumor-producing homologue. Such genes are also called proto-oncogenes and can be activated by mutation, amplification, or overexpression to become cancer producing.

condensation The chemical reaction that results in the joining of two molecules with the elimination of a water molecule. An example is the formation of the peptide bond between two amino acids by the reaction of the amino group of one amino acid with the carboxyl group of the other.

condensing vacuole A membrane-bound vacuole arising from the Golgi complex (see GOLGI APPARATUS) and developing into a secretory granule by the progressive loss of water.

conditional mutation A mutation expressed only under certain conditions. An example is a temperature-sensitive mutation that encodes a protein that is functional at certain permissive temperatures (e.g., 32°C), but not functional at higher nonpermissive temperatures (e.g., 42°C). Such mutations can define essential genes, because mutations in essential functions are lethal events, but conditional mutations allow the organisms to survive at permissive temperatures.

conformation The three-dimensional structure of a macromolecule, such as a protein.

congenital Acquired during development in the uterus.

Bacterial Conjugation

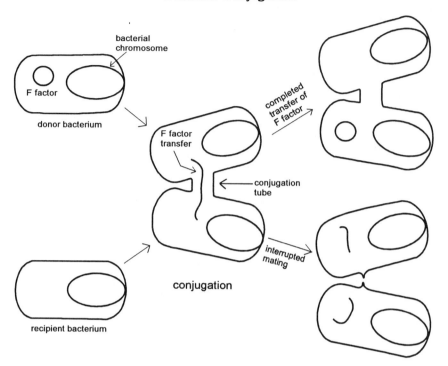

conidiophore A specialized structure at the tips of the hyphae in fungi from which the fungal spores (known as conidia) bud.

conjugation A means of gene transmission between any bacterial strain that carries an F factor (small episomal DNA segment) carried either on the bacterial chromosome or extrachromosomally, and another strain that lacks the F factor. During conjugation neighboring bacteria come into direct contact with one another and transfer DNA by means of a mating tube which forms at the point of contact. Because the genes from the donor are always transferred in a given order, conjugation has been used to map genes on the bacterial genome by observing which genes are transferred to recipients following controlled interruption of the mating process. See HIGH FREQUENCY RECOMBINATION STRAIN.

connective tissue FIBROBLAST cells that secrete collagen to give cells the adhesive strength needed to maintain form. Some examples of connective tissue are bone, cartilage, tendons, and ligaments.

connexon A structure of the GAP JUNCTION composed of six protein subunits around a hollow center. Two aligned connexons of two cells provide a means of communication between the two cells.

consensus sequence An order of bases with the most common NUCLEOTIDE at each position when different examples are compared. Consensus sequences are found in PROMOTERs and are responsible for binding RNA polymerase and other proteins needed for transcription. Consensus sequences also signal other events, such as splicing of INTRONs out of primary transcripts.

conservative replication A mechanism or DNA replication in which each strand of a parental molecule remains together after replication. (See SEMIDISCONTINUOUS REPLICATION.)

constant region The carboxy terminal region of the heavy chain or light chain of the antibody molecule that has the same or nearly identical amino acid composition as each member of the same class, such as IgM, IgA, IgE, IgG.

constitutive gene Genes expressed continuously and not subject to INDUCTION or repression. Such genes encode housekeeping functions and are expressed in all cells at a low level.

constitutiveheterochromatin Chromosomal regions that remain in a permanently condensed state during interphase in every cell in the organism and are never genetically active in any cell, such as CENTROMERES.

constitutive mutant An organism with a mutation in some regulatory gene so that the expression of the gene(s) it controls is constitutively expressed. (See CONSTITUTIVE GENE.)

contact inhibition The property of normal animal cells in culture to stop dividing once they have formed a contiguous monolayer over the surface of the medium on which they are growing.

contamination Growth of undesirable organisms in some culture or material.

continuous culture A system that uses a chemostat or turbidostat to maintain a cell culture at a steady growth rate.

contractile ring A beltlike structure of actin microfilaments under the plasma membrane that functions to divide an animal cell into two daughter cells after mitosis.

controlling element Transposable elements that cause mutations and chromosomal breakage when they are transposed into, or excised out of, genes.

Coombs' reaction An immunological test for the presence of antibodies that react with red blood cells (RBCs); such autoantibodies are present in the blood of individuals with certain hemolytic diseases. The test involves reacting blood cells first with the serum from individuals to be tested and then with antiantibodies to the antibodies that may be bound to the RBCs. If autoantibodies to RBCs are present, then the RBCs used in the Coombs' test will be observed to form large clumps.

coordinated enzyme synthesis Regulation of the synthesis of enzymes involved in the same metabolic process by the same event or signal. Generally accomplished in bacteria by the organization of the genes that encode the enzymes into operons with a single regulatory element. In higher organisms, the genes are usually scattered but have common regulatory elements that respond to the same signal.

coordinate regulation Expression of multiple genes in unison (e.g., the genes of an operon). (See LAC OPERON.)

copia elements A family of transposable elements in *Drosophila*. A typical *Drosophila* genome carries about 50 such elements in widely scattered regions.

copolymer A synthetic polymer of two deoxyribonucleotides in random order (e.g., ACACCACCCAA) or two ribonucleotides in alternating order (e.g., AUAUAUAUAU). Copolymers were used to elucidate the genetic code by inserting them in an *in vitro* protein synthesizing system and analyzing the amino acid composition of the resulting polypeptides.

copy choice   A mechanism of genetic recombination in which the recombinant molecule is formed by selectively replicating parts of the parental DNA molecules.

copy number   The number of PLAS-MID molecules per bacterial cell. Some plasmids are said to be relaxed in the control of their replication and are defined as high-copy-number plasmids (e.g., 20 to 100 copies per cell). They are used as CLONING VECTORS, and result in high yields of recombinant DNA or the protein encoded by the recombinant. Stringently controlled plasmids exist in cells in low copy number, or one to a few copies per cell. These plasmids are used to clone genes that produce proteins toxic to bacterial cells when produced in high concentrations.

cordycepin   An antibiotic that acts by blocking transcription, it is a derivative of the normal nucleoside, adenosine, in which the hydroxyl group on the 3'-carbon is missing. When cordycepin becomes incorporated into newly synthesized RNA in place of the normal adenine nucleoside, the RNA strand terminates.

core particle   An octamer of HIS-TONES (H2A, H2B, H3, and H4) with 146 bp of DNA wrapped $1^3/4$ times around in a NUCLEOSOME.

corepressor   The EFFECTOR molecule that binds to a repressor to form a complex. The effector–corepressor complex functions to repress or prevent transcription of a bacterial operon. (See LAC OPERON.)

cornybacteria   Small, Gram-positive, straight to slightly curved, rod-shaped or frequently club shaped bacteria, with aerobic to facultative heterotrophic metabolism. A pathogenic member of this group is *cornybacterium diphtheriae,* the causative agent of diphtheria.

corpus luteum   The temporary endocrine gland formed from a ruptured ovarian follicle after release of an egg.

cortical cytoplasm   A region of the egg cytoplasm just under the cell membrane that undergoes rearrangements after fertilization and has profound consequences in embryo development.

cortical reaction   Release of enzymes from the cortical vesicles after fertilization of an animal egg that results in a hardening of the vitrelline membrane to prevent additional sperm penetration.

cortical vesicles   Membrane-bound structures of the egg cell that release proteases and other enzymes during the process of fertilization.

corticosteroid   The steroid hormone, a derivative of cholesterol, that is synthesized in the adrenal cortex.

cos cells   A derivative of CV-1 monkey cells that are infected with, but do not produce, SV40 virus. The cos cells express the SV40 EARLY GENES (T antigens) needed for viral replication and are used as host cells in cloning experiments when derivatives of SV40, lacking the early genes, are used as CLONING VECTORS for eukaryotic DNA. The early genes expressed by the cos cells allow the recombinant molecules to replicate, and thus amplify its cloned inserts.

cosmid   A specialized cloning vector constructed from the following genetic elements: a plasmid origin of replication, an antibiotic resistance gene, and the cos sites of λ DNA. These molecules can be packaged in the laboratory into a normal λ bacteriophage capsid and can then be transferred into host cells by the normal process of viral infection.

Because they have a plasmid origin of replication, cosmids can be grown in bacteria until they are to be used for cloning. Cosmid vectors can be used to clone DNA fragments up to 47 kilobases in length.

cos site A single-stranded segment, 12 nucleotide bases in length, that protrudes from the 5'-ends bacteriophage λ DNA (5'-overhangs). Because the segments at each end of the λ DNA are complementary to one another, the DNA becomes circularized after entering a host cell by complementary base pairing of the ends to one another. During the life cycle of the bacteriophage, the cos sites serve as markers for packaging the λ bacteriophage DNA into CAPSIDS during the later stages of maturation of the progeny phage particles.

cotransduction The introduction of two linked genes into a bacterial cell by the genetic transmission process of transduction.

cotransformation The introduction of two linked genes into a bacterial cell on the same fragment of DNA by the genetic transmission process of transformation.

cotranslational transfer The insertion of one end of a polypeptide into the membrane of the ENDOPLASMIC RETICULUM before synthesis of the whole polypeptide is completed. (See LEADER SEQUENCE.)

cotransport The simultaneous movement across a membrane of two different substances in a coupled manner. The two substances may move in the same direction (symport) or in the opposite direction (antiport).

Coulter counter An instrument that automatically counts cells by measuring the changes in resistance that occur when cells in suspension are passed through a small slit.

coupled reactions Two enzymatically controlled chemical reactions that must occur simultaneously. For example, many reactions that require an input of energy to proceed are coupled to the hydrolysis of ATP (ATP→ADP+Pi), which releases 7 kcal of energy.

coupled transport The obligatory simultaneous transport of two solutes across the cell membrane. (See COTRANSPORT.)

covalent bonds Strong chemical bonds between atoms involving the sharing of two or more electrons.

covalently closed circular (ccc) DNA A circular double-stranded molecule of DNA, such as a plasmid, in which there are no nicks or breaks in the sugar–phosphate backbone. Usually, ccc DNA exists as supercoiled—that is, the molecule folds in on itself, due to strain in the molecule. If a nick is introduced into the backbone, the molecule relaxes and is referred to as an open circle (oc). (See SUPERCOILED DNA.)

coxsackie viruses An antigenically distinct group of viruses of the enterovirus genus (viruses found in the intestines and excreted in the feces), including some human pathogens.

CpG-rich islands Regions of DNA believed to be regulatory elements of gene activity characterized by an unusually high content of cytosine and guanine nucleotides arranged in the repeating sequence CGCGCGC. . . . CpG islands are located in a segment 5' to the coding regions in many genes and are often sites of methylation.

creatine phosphate A high-energy compound in muscle cells used to regenerate the ATP needed for muscle contraction.

Crick, Francis (b. 1916) English scientist who, with James Watson, won the Nobel Prize for physiology or medicine in 1962 for postulating a double-stranded helical stucture for DNA, using the x-ray diffraction data of Maurice Wilkins, also a Nobel Prize winner in the same category in 1962, and Rosalind Franklin. The double helix accounted for the known physical and chemical properties of DNA, but also suggested a mechanism for its replication.

crista The infolding of the inner membrane of the mitochondrion, which increases the surface area of the membrane responsible for electron transport and production of ATP via oxidative phosphorylation.

critical concentration The minimal concentration of subunits required for formation of a biopolymer. For example, the formation of microtubules begins to occur in the cytoplasm of cells only after the concentration of the tubulin subunit reaches a critical concentration.

critical dissolved oxygen concentration (Ccrit) The concentration of dissolved oxygen in a submerged culture when oxygen is the limiting substrate. The air supply to a fermentor is adjusted to maintain an oxygen level above its Ccrit.

cro protein A protein that interferes with the synthesis and action of the C repressor of λ BACTERIOPHAGE and is necessary in the lytic response of the phage after infection into a cell. (See LYSE.)

crossing over The physical exchange of genetic information between a pair of homologous DNA molecules.

cross-linking A reaction in which two strands of DNA are covalently bonded together. Certain mutagenic agents, such as X-rays, cause cross-linking, and the DNA must be repaired if it is to replicate and function properly.

crossover fixation An alternative model to SALTATORY REPLICATION to explain the occurrence of highly repeated sequence. In crossover fixation, additional copies of a certain sequence are created on one DNA strand by unequal crossing over.

cross-reactive antibodies Nonspecific antibodies that will bind to antigens and give a false positive response in an antigen-antibody test.

crown gall plasmids The Ti (tumor inducing) PLASMID of *Agrobacterium tumefaciens* responsible for the malignant transformation of dicotyledonous plants infected with this organism. Part of the plasmid DNA incorporates into the plant chromosome to cause the production of a tumor. Those plasmids lacking the tumor-producing genes have been constructed as potential vectors for recombinant DNA molecules for plant genetic engineering.

crown gall tumor See CROWN GALL PLASMIDS.

cruciform structure DNA structure in which strands separate and self-anneal through complementary base pairing to form cruciforms or crosslike structures. Cruciforms can arise at regions of inverted base-pair repeats.

cryogenics The science of freezing, especially with reference to methods for producing very low temperatures.

cryopreservation The preservation of cells, organs, tissues, or other biological materials at very low temperatures in freezers ($-80°C$), over dry ice ($-79°C$), or in liquid nitrogen ($-196°C$). At low temperatures, preserved biological materials remain genetically stable and metabolically inert.

cryoprotectants Chemicals that reduce the formation of ice crystals during freezing so that survival of cryopreserved cells is enhanced. Common cryoprotectants are dimethyl sulfoxide (DMSO), glycerol, and sucrose.

Cruciform DNA

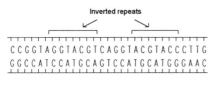

Inverted repeats

```
CCGGTAGGTACGTCAGGTACGTACCCTTG
GGCCATCCATGCAGTCCATGCATGGGAAC
```

intrastrand base paring
(hairpin formation)

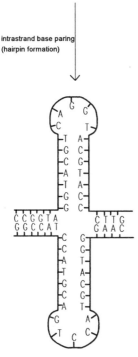

cryptic plasmid A plasmid containing no genes or apparent phenotypic markers other than those needed for replication and transfer.

crystallography (X-ray) A technique used to analyze the structure of molecules by analysis of diffraction patterns of X-rays that pass through crystal specimens.

C_0t value A parameter (molar concentration of DNA multiplied by time) used to plot the reassociation of denatured DNA.

culture A population of cells cultivated in a medium.

curing Any action that causes the loss of a plasmid or lysogenic bacteriophage from a culture of bacteria.

cutaneous Pertaining to, existing on, or affecting the skin.

cuvette The sample container for a spectrophotometer or other instrument used to make measurements on liquid samples.

c value A value representing the total amount of DNA, given in base pairs, in the haploid genome of a species.

Cyanobacteria Blue-green algae. Prokaryotic, photosynthetic, oxygen-evolving organisms.

cyanogen bromide (CNBr) A chemical that recognizes methionine residues and cleaves polypeptide chains at these residues. CNBr is used to cleave genetically engineered proteins that have been constructed as a composite of cloned material and vector material. It is also used to cross-link proteins to various support materials for affinity chromatography purposes.

cyclic adenosine monophosphate (cAMP) A molecule of adenosine mo-

Cyclic AMP (3′-, 5′-AMP)

nophosphate in which there is a covalent bond between the 3'-hydroxyl (OH) and the 5'-phosphate group. It is an important molecule in controlling metabolic processes in higher organisms (see CYCLIC AMP-DEPENDENT PROTEIN KINASES). Since its intracellular concentration is often controlled by hormonal action, and the metabolic activities it controls is in response to the hormone, it is called a "second message." cAMP also plays a role in the control of bacterial metabolism by binding with the CAP protein and turning on transcription.

cyclic AMP-dependent protein kinases Enzymes that add phosphate groups to proteins either at a serine or tyrosine residue. Phosphorylation of proteins is an important mechanism of regulation of metabolism in higher organisms, because the added phosphate group activates or inactivates the protein, and thus stimulates or inhibits the metabolic reaction.

cyclic guanosine monophosphate (GMP) A molecule of guanosine monophosphate in which there is a covalent bond between the 3'-hydroxyl (OH) and the 5'-phosphate group.

cyclin A protein with an action similar to MPF (See M PHASE)—that is, to induce entry of cells at any stage of the CELL CYCLE into mitosis (M phase). Cyclin, like MPF, was isolated from dividing cells in the embryo of the frog, *Xenopus laevis,* and is made at all phases of the cell cycle but is destroyed during mitosis.

cyclohexamide An antibiotic used as an agricultural fungicide, that inhibits yeasts and other fungi, but not bacteria.

cycloserine An antibiotic from *Streptomyces* that acts by blocking two steps in the biochemical pathway by which the bacterial cell wall is synthesized. Because cycloserine is structurally similar to the amino acid D-alanine, it competitively inhibits the incorporation of D-alanine into a pentapeptide used to construct the bacterial cell wall.

cysteine An amino acid with a sulfhydryl group in its side chain:

$$CH_2-HS-CH_2-SH$$

Disulfide bonds between two cysteine residues on the same polypeptide chain or between chains contribute to the overall shape of the protein.

cystic fibrosis An inherited disease afflicting almost 1 in 2000 children in the United States. In 1989 the gene, whose mutant allele accounts for a majority of the cases, was cloned.

cystic fibrosis transmembrane conductance regulator (CFTCR) protein The product of the gene found to be defective in patients with cystic fibrosis. It is believed to be an integral membrane protein that serves as a channel for transporting certain ions into and out of the cell.

cytidine monophosphate (CMP) The nitrogenous base CYTOSINE attached to a ribose sugar molecule with a phosphate residue at the 5'-end of the ribose. Cytidine triphosphate (CTP) has three phosphate residues.

cytochalasins A family of drugs produced by certain fungi that interferes with polymerization of actin microfilaments, and hence inhibits cell movements depending on actin polymerization–depolymerization reactions.

cytochrome c oxidase An enzyme complex of the electron transport chain that reduces molecular oxygen to water.

cytochrome P-450 (P450) A cytochrome, found in the smooth ENDOPLASMIC RETICULUM, important in drug detoxification, especially in the liver. It is a type of mixed-function oxidase that carries out hydroxylation reactions (addition of OH groups) to molecules, thus aiding in solubilizing them so that they can be flushed from the body.

cytochromes Heme-containing proteins of the electron transport chain that carry out oxidation–reduction reactions, thus passing electrons down the chain from iron atom of one cytochrome to iron atom of another until the final electron acceptor, molecular oxygen.

cytokinesis The process of dividing the ctyoplasm of a cell into two daughter cells, following mitosis.

cytokinins Substances that promote cell division and cell and shoot differentiation in plant tissue cultures. Some common cytokinins are benzylaminopurine (BAP) and 2-isopentenyladenine.

cytology The study of cells based on microscopic observations.

cytoplasm The liquid colloidal substance between the cell membrane and nucleus of the cell.

cytoplasmic   inheritance The genes in mitochondria or chloroplasts found in the cytoplasm of the ovum and referred to as maternal inheritance.

cytoplasmic streaming The back-and-forth movement of cytoplasm in some algae and the circular flow of cytoplasm around a central vacuole in a plant

Cytosine

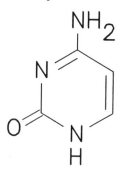

cell, also known as cyclosis.

cytosine A pyrimidine, it is one of the nitrogenous bases in nucleic acids that forms hydrogen bonds with the purine guanine.

cytoskeleton A complex network of microtubules, microfilaments, and intermediate filaments extending throughout the cytoplasm that gives shape to a eukaryotic cell and is involved in cellular movement.

cytosol The cytoplasm that contains the organelles of a eukaryotic cell.

cytotoxic T cell (Tc) An activated T lymphocyte, also known as a killer T cell, that LYSES cells recognized as a combination of self and foreign, such as virally infected cells, tumor cells, and foreign tissue grafts.

D

dalton A unit of mass, used generally for macromolecules, equal to 1.000 on the atomic mass scale, about the same as that of a hydrogen atom. Dalton can be used interchangeably with molecular weight. Thus, a 100,000-dalton (or 100-

kilodalton) protein can be described as having a molecular weight of 100,000.

dansyl chloride A compound that reacts with the amino group of an amino acid to produce a fluorescent derivative that can be easily detected and identified. It is used in procedures to identify the amino terminal residue of peptides.

dark reactions   A series of enzymatically catalyzed reactions in which organisms carrying out photosynthesis synthesize organic compounds in the form of sugars from inorganic carbon dioxide. These reactions use energy, in the form of ATP, and reducing power, in the form of NADPH, made during the light-phase reactions of photosynthesis.

deaminase   An enzyme that removes amino groups from molecules.

deamination   The process by which a deaminase removes amino groups from molecules. Deamination of bases in DNA results in mutations, and cytosine is the most susceptible base.

death phase   The final phase in the growth curve of a population of cells in which the cells die exponentially. That is, for each time increment a certain percentage of cells dies.

decline phase   See DEATH PHASE.

defective virus   A virus that is missing some essential genetic information so that it cannot reproduce itself. Such viruses can be propagated in a host cell only if a helper virus, which supplies the missing proteins, coinfects the same host cell.

defined medium   A medium used to grow organisms in which all the components are known. For heterotrophs, that would be a medium with a known carbon source, nitrogen source, metals, and any amino acids, vitamins, or other growth factors required by the organism.

degenerate code   Referring to the fact that in the GENETIC CODE many amino acids are specified by more than one CODON or sequence of three bases (triplet). The degeneracy of the code accounts for 20 different amino acids encoded by 64 possible triplet sequences of four different bases (see NUCLEIC ACIDS). For example, the amino acid leucine has six codons: UUA, UUG, CUU, CUC, CUA, CUG. (See WOBBLE.)

degradation   The process by which substances are broken down. A degradative pathway is one in which molecules are enzymatically cleaved into smaller molecules.

dehydration-condensation reaction   The joining of two molecules with the elimination of a molecule of water. (See HYDROLYSIS.)

dehydrogenation   The process by which hydrogen ions or protons are removed from an organic molecule. It is carried out by enzymes called dehydrogenases. Also called oxidation.

delayed hypersensitivity   An allergic reaction that takes 24 to 48 hours to appear. An example is the skin test for exposure to tuberculosis. After the ALLERGEN is injected, a positive response (a swelling at the injection site) does not appear before 48 hours.

deletion mutation   A change on the DNA due to the elimination of one or more nucleotides. A deletion can alter the genetic information in a very profound way. The deletion of one base pair results in a FRAMESHIFT, where every codon is changed following the deletion; a deletion of many bases results in a message with fewer CODONS.

delivery system   An artificial system to deliver a drug to a specific target, such as inclusion of a drug in a LIPOSOME or conjugating a drug to an antibody. (See FUSOGENIC VESICLE.)

demyelination   The loss of the MYELIN sheath, which consists of layers of membrane that surround segments of nerves and provide rapid transmission of nerve impulses down such nerves. Demyelination occurs in some degenera-

DNA Denaturation

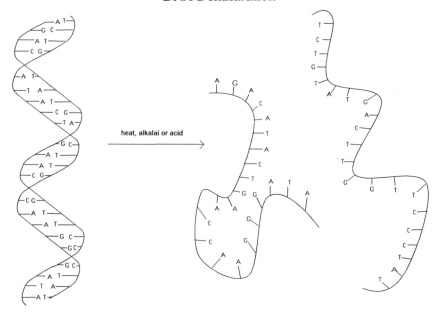

heat, alkalai or acid

tive nerve diseases, such as multiple sclerosis and polio, resulting in loss of function of those demyelinated nerves.

denaturation Change in the three-dimensional shape or structure of a protein or nucleic acid by a physical or chemical agent such as heat or strong acid (the denaturant) such that normal functioning is altered.

denaturation of DNA Splitting of the double-stranded structure into single strands by heating the molecule or treating it with acid, alkali, salts, or urea.

dendrite A branch of a nerve cell that receives signals and transmits them inward toward the nerve cell body.

Denhardt's solution A commonly used solution for carrying out PROBE hybridizations on filters (e.g., SOUTHERN BLOT). Denhardt's solution contains polyvinylpyrrolidone (PVP), ficoll, bovine serum albumin, and a nonspecific DNA

at high concentration to prevent nonspecific probe hybridization.

denitrification The process of reducing nitrogen compounds to a lower oxidation level, for example, nitro (NO_3) to nitrate (NO_2) or nitrate to nitrite (NO).

density gradient A solution in which there is a range of densities with the solute being more concentrated at the bottom and less so at the top. These gradients can be stepwise (formed by discrete layers of different density solutions) or continuous (formed by small incremental changes in density). The gradients are generally made from solutions of sucrose or the heavy salts cesium chloride and cesium sulfate.

density gradient centrifugation Technique used to separate macromolecules according to their buoyant densities or molecular weights by either layering the macromolecules on top of a preformed gradient and subjecting the gradi-

Density Gradient Centrifugation

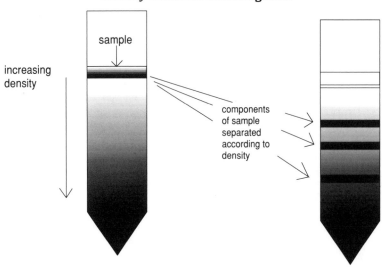

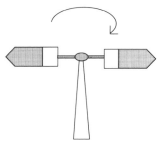

centrifugation

ent to centrifugation or by mixing the molecules in a solution of some support that will form a gradient upon centrifugation. (See CENTRIFUGE, CESIUM CHLORIDE (CsCl) DENSITY CENTRIFUGATION, DENSITY GRADIENT.)

deoxynucleoside Any nitrogenous base found in nucleic acids (adenine, guanine, thymine, uracil, or cytosine)

attached to deoxyribose, a 5-carbon sugar. (See NUCLEOSIDE.)

deoxyribonuclease (DNase) An enzyme that breaks the chemical bond between the phosphate and sugar groups (the backbone) of DNA molecules. These enzymes can be exonucleases, which remove deoxynucleotides from the ends of the mole-

Deoxyribose

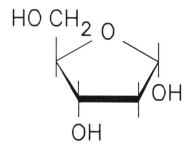

cule, or endonucleases, which cleave bonds of internal nucleotides.

deoxyribonucleic acid (DNA) A macromolecule consisting of two complementary (see COMPLEMENTARY BASE PAIRING) chains of deoxyribonucleotides. The chains are formed by chemical bonds between the sugar and phosphate portions of the deoxyribonucleotides, and the two chains are held together by hydrogen bonds between the bases (A pairs with T, and G pairs with C). DNA contains the genetic information of the cell, because the sequence of nucleotides of the chains specifies the sequence of amino acids of proteins made by the cell.

deoxyribonucleotide A deoxyribonucleoside (see DNA) with phosphate groups attached to the sugar, usually at the 5'-position, the basic building block of DNA. (See NUCLEOTIDE.)

depolarization The change in electrical charge across a membrane. When a nerve cell receives an impulse, it becomes momentarily depolarized; its interior becomes more positively charged with respect to its exterior. Repolarization restores its interior to a negative charge. A nerve impulse is propagated down a nerve fiber by waves of depolarization–repolarization events. (See ACTION POTENTIAL.)

depurination Removal of purines (guanine or cytosine) from DNA with the sugar-phosphate backbone remaining intact. Such a process occurs either enzymatically, as in a DNA repair process, or nonenzymatically, such as when a chemical interacts with the base, and weakens its bond to the sugar residue.

desalting Removal of salt. Accomplished for preparations of macromolecules by DIALYSIS or column chromatography. (See GEL FILTRATION.)

desmin A protein component of the INTERMEDIATE FILAMENTS found in muscle cells.

desmoplaquin One of the protein components of the DESMOSOME.

desmosome Region of tight contact between adjacent epithelial cells that strengthens tissues and enables tissue cells to function together. Belt desmosomes are bands of attachment that encircle the cell, and spot desmosomes are small local points of contact.

desmotubule A tubular structure that lies in a channel of the plasmadesmata, a means of communication between two plant cells. Such a channel is made up of the fusion of cell membranes from two adacent cells through the pores of the cell wall.

desulfurization The removal of sulfur from a molecule.

detergent A molecule with a hydrophobic part and a hyrophilic portion that can dissolve lipids (fats and oils).

determination The irreversible commitment of a cell to a particular developmental pathway. If a determined cell is transplanted, it will develop into the same structure as if it had not been transplanted.

Deuteromycetes One of the four major classes of fungi (also called *Fungi imperfecti*), important because it contains the majority of human pathogens.

dextran A storage polysaccharide in yeasts and bacteria made up of glucose.

dextranase   An enzyme that catalyzes the breakdown of dextran.

dextrin   A dextrin, a molecule made up of several glucose residues. It is one of the products resulting from hydrolysis of starch by α-amylase.

dextrose   Another name for D-glucose, the most common sugar in living organisms.

dextrotatory isomer   An optical isomer of a sugar that rotates polarized light to the right. Dextrose is the dextrotatory isomer of glucose.

diacylglycerol (DAG)   A molecule of glycerol with two fatty acids attached to it by ester linkages. It is formed along with INOSITOL TRIPHOSPHATE (InsP3) by hydrolysis of a membrane lipid (phosphatidylinositol-4, 5-bisphosphate) by the enzyme PHOSPHOLIPASE C. Both DAG and InsP3 serve as second messengers in the cell. DAG activates PROTEIN KINASE C, which activates other enzymes.

diagnostic   (1) A test used to determine the source of a problem. (2) The method of determining the nature of a disease by analyzing the symptoms. (3) A specific characteristic that allows one to determine the source of a problem or the nature of a disease.

diagnostic test   A procedure to determine the nature of a problem. (See DIAGNOSTIC.)

dialysis   (1) A technique to separate molecules based on their permeability through a semipermeable membrane. The membrane allows water and small molecules, such as salts, to pass through while retaining large molecules, such as proteins. Thus, proteins can be desalted by dialysis. (2) A medical procedure used to clear the blood of impurities after kidney failure. (See DESALTING.)

diatomaceous earth   A finely pulverized mixture of earth composed largely of the silicon shells of the microorganism's diatoms. Used as a filtering substance or as an absorbant.

dibasic   An acid with two hydrogen atoms that may be replaced by basic molecules or metal ions to form a salt.

dicentric chromosome   A chromosomal aberation involving breakage and then fusion of chromosomal fragments, resulting in the formation of a hybrid chromosome with two CENTROMERES.

dicotyledon   Any plant characterized by flower parts in fours and fives, netveined leaves, a cambium, and an embryo with two cotyledons, or two seed leaves.

dictysome   A stack of flattened membranous sacks in plant cells located adjacent to the ENDOPLASMIC RETICULUM. Its function is to mediate the secretion of proteins outside the cell or to target newly synthesized proteins to organelles such as the lysosome (called the GOLGI APPARATUS in animal cells).

dideoxynucleotide   A nucleotide with a ribose having a hydrogen atom at the 3'-position instead of an OH group. Such a nucleotide cannot form a 3'- to 5'-phosphodiester linkage (the linkage of the sugar-phosphate backbone of DNA) with another nucleotide. Thus, adding a dideoxynucleotide to a growing DNA chain will terminate further synthesis of the chain. (See DIDEOXY SEQUENCING.)

dideoxyribonucleotide triphosphate (ddNTP)   See SANGER SEQUENCING.

dideoxy sequencing   An enzymatic method, developed by Fred Sanger, of sequencing DNA using dideoxynucleotides to prematurely stop the synthesis of DNA chains at specific points. DNA to be sequenced is divided into four tubes containing DNA polymerase, the four deoxynucleotides (dA, dT, dG, dC), and one of the dideoxynucleotides ddA, ddT, ddG, or ddC). The ratio of dideoxynucleotide to regular nucleotide is fixed, so during DNA synthesis the DNA polymerase has the option of incorporating a regular or a dideoxynucleotide. Since incorporation of a DIDEOXYNUCLEOTIDE into a growing DNA chain stops further synthesis of that chain, each tube contains a series of fragments, each ending with the dideoxynucleotide of that tube; for example, the tube containing ddA has fragments ending in A. The size of each fragment can be determined by gel electrophoresis, and the sequence can be read upon the gel. For example, if the tube containing ddA produces three fragments, with sizes of 4, 7, and 10 bases, the sequence of the DNA synthesized has an A residue at the 4th, 7th, and 10th positions. (See SANGER SEQUENCING.)

differential centrifugation   A technique to separate cells, organelles, or molecules differing in size or density by using successively higher centrifugal forces.

differentiation   The process during the development of an embryo in which cells become specialized in structure and function and go on to form different tissues of the adult.

differentiation antigen   Any biomolecule detectable by an immunologic assay only in a specific cell subtype in an organism and may therefore be used as a marker of that subtype.

diffusion   Free movement of molecules from an area of greater concentration to one of lower concentration.

diffusion coefficient or constant (k_D) The measure of the ability of a solute to diffuse through a concentration gradient. The factors affecting k_D include particle size, degree of polarity, and temperature. Diffusion rate depends on k_D as follows: $v=k_D$ ([X]$_{outside}$ − [x]$_{inside}$), where v is the diffusion rate and [X]$_{outside}$

of solute [X] of the concentration gradient.

digitalis   Medicine made from dried leaves of purple foxglove *(Digitalis purpurea)* and used to stimulate the heart.

digoxigenin   A plant-derived steroid that, when covalently bound to a biological PROBE molecule, has been used as a HAPTEN in some antibody–hapten-based probe systems.

dihydrouridine   An unusual pyrimidine base, found only in tRNA, derived from uridine by adding two hydrogen atoms.

dimer   (1) A molecule with two subunits, which may or may not be identical. (2) Denoting two units (e.g., a dimer of nucleotides, dCdG).

dimethoxytrityl (DMT)   A molecule used as a blocking group to prevent unwanted reactions in automated oligonucleotide synthesis.

dimethyl   sulfoxide   (DMSO) $[(CH_3)_2SO]$.   A reagent used for the cryopreservation of cultured animal cells and to increase the efficiency of TRANSFECTION of DNA. (See CRYOPROTECTANTS.)

dimorphism   The state of having two different forms. In botany, it describes a plant or a species of plant that has two distinct leaf types, or flowers, or some other structure. In zoology, it describes two individuals of the same species exhibiting different coloring, size, or other characteristics.

dioxin   A group of heterocyclic hydrocarbons or any of a number of isomers of the chlorinated teratogen, TCDD, that is highly toxic and found as impurities in some defoliants and herbicides.

diploid   Having two sets of chromosomes, so each gene is represented twice in a cell or organism. Describing a cell

The Water Dipole

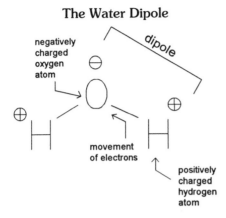

negatively charged oxygen atom

dipole

movement of electrons

positively charged hydrogen atom

or organism containing two copies of each chromosome. (See HAPLOID.)

diplotene   A stage of prophase I during MEIOSIS in the formation of germ cells. During diplotene, the chiasmata, or region where crossing over took place, can be visualized.

dipole   A polar molecule in which the centers of positive and negative charge are separated. A molecule of water has a triangular shape and exists as a dipole. The oxygen at the head of the triangle is electronegative, and the two hydrogen tails are electropositive. (See POLARITY.)

direct terminal repeats   Sequences of nucleotides duplicated on each end of a polynucleotide molecule. (See LONG TERMINAL REPEAT, TERMINAL REDUNDANCY.)

disaccharide   A molecule consisting of any two sugar units. Maltose is a disaccharide consisting of two glucose molecules linked by an α-glycosidic bond. Sucrose is composed of fructose and glucose linked by an α-glycosidic bond.

disc   electrophoresis   Shortened term for discontinuous electrophoresis, a refinement of polyacrylamide GEL ELECTROPHORESIS in which the sample is electrophoresed through two polyacrylamide phases: a low-percentage (stacking) gel that sits atop a higher-percentage (resolving) gel. The two-phase approach produces higher resolution between closely migrating bands.

disinfectant   Any chemical that can kill bacteria and viruses.

dissociation   constant   The constant that quantifies the dissociation of two atoms, molecules, or even large particles, from one another. For the dissociation of substance A from substance B, where AB is a complex of A and B, AB $\longrightarrow$ A+B, $K_{dissociation}$=[A][B]/[AB], where [ ] represents molar concentration.

distillation   The process of separating and purifying liquids from a mixture based on each liquid's boiling temperature. The more volatile substance boils at a temperature lower than the others in a mixture. The vapor is then collected, cooled, and condensed, thus extracting and refining it from the mixture.

disulfide bond   A covalent bond between two sulfhydryl (SH) groups by oxidation to form an S—S linkage. Such bonds occur in proteins between cysteine residues and stabilize the TERTIARY STRUCTURE of the protein.

divalent   An atom or radical group having two valences or the ability to combine with two different atoms or molecules.

D loop   A structure of DNA in which there is a localized denaturation of the duplex or displacement of a single strand from the duplex, resulting in a shape that resembles the letter D. This structure is usually stabilized with proteins called single-stranded binding proteins.

DNA,   repetitive   Sequences repeated many times on the genome.

DNA Polymerase

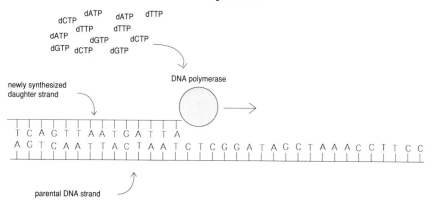

These sequences vary in length, from 3 to 5 to 300 bp, and are found on the genome in hundreds to thousands of copies. Some of the repetitive DNA makes up the satellite DNA, a distinct band from the bulk of chromosomal DNA found after cesium chloride density centrifugation. (See ALU ELEMENTS.)

DNA cloning Any procedure that generates many copies of a particular DNA sequence. The sequence can be inserted into a PLASMID or BACTERIO-PHAGE, which will be duplicated many-fold in a bacterial cell, or the sequence can be copied manyfold by POLYMERASE CHAIN REACTION (PCR).

DNA glycosylase An enzyme that recognizes a deaminated base and cata-lyzes its removal from the DNA mole-cule, creating an apurinic or apyrimidinic site in the DNA molecule.

DnaG primase The enzyme respon-sible for catalyzing the formation of the short RNA PRIMERS in OKAZAKI FRAG-MENTS.

DNA gyrase An enzyme that cata-lyzes the introduction of negative su-percoils or relaxes positively supercoiled DNA by unwinding one strand of duplex DNA around the other, so that each

strand is wrapped around the other less than one turn for every 10 bases.

DNA photolyase An enzyme that catalyzes the repair of pyrimidine dimers formed as the result of ultraviolet irradia-tion. The first step in the repair involves the excision of the dimer, which occurs only in the presence of visible light. DNA photolyase is also known as photoreacti-vating enzyme. (See ULTRAVIOLET RE-PAIR.)

DNA polymerase I A specific DNA polymerase that has not only the 5' →3' polymerizing activity but also two nucleo-lytic or degradative activities: a 3' →5' exonuclease (an editing function) and a 5' →3' exonuclease. This enzyme, with all of its activities, is used by the cell during different steps of DNA replication and DNA repair processes. In addition, purified DNA polymerase I, with or with-out its nuclease activities, is used in vari-ous in vitro procedures, such as preparing labeled DNA probes and DNA sequencing via the dideoxy method. (See KORNBERG ENZYME.)

DNA polymerase(s) Any enzyme that can use a chain or strand of deoxy-nucleotides as a template and synthesize a complementary strand. All DNA poly-merases synthesize DNA from the 5'-

phosphorylated end to the 3'-hydroxyl end.

DNA probe   A sequence of deoxynucleotides used to identify or isolate specific genes or RNA transcripts that have complementary sequences. Such probes are used in hybridization procedures (Southern, northern, slot, and dot blots, and colony or plaque hybridizations) and are labeled with either a radioactive atom of ^{32}P or ^{35}S, which allows detection by autoradiography, or nonradioactive materials, such as biotin or digoxigenin, which are detected via specific reactions. (See PROBE.)

DNA repair   Any process that restores damaged DNA. Generally the processes are multistep, requiring an enzyme to remove the damaged nucleotide (see DNA GLYCOSYLASE, ENDONUCLEASE) alone or with other nucleotides, a polymerase (see DNA POLYMERASE I) to replace the removed nucleotides and an enzyme to seal the sugar–phosphate backbone. (See DNA LIGASE, EXCISION REPAIR.)

DNA–RNA hybrid   A DNA–RNA duplex molecule composed of a single chain of deoxyribonucleotides (DNA) and a chain of complementary ribonucleotides (RNA). Such molecules may be created experimentally from purified DNA and RNA or when chromosomal DNA is fragmented, heated, and mixed with RNA transcripts.

DNase I   A DNA-degrading enzyme that catalyzes the cleavage of phosphodiester bonds of DNA; DNase I is isolated in large quantities from pancreas. (See ENDONUCLEASE.)

DNase I hypersensitivity sites   Regions on the chromosome that are extremely sensitive to digestion by DNase I and generally found near active genes where transcription factors or other regulatory elements bind to the DNA. (See HYPERSENSITIVE SITE.)

DNase I sensitivity   Increased susceptibility to digestion by DNase I correlates with genes that are actively transcribing RNA. This shows that the chromatin of genes being expressed has an open conformation accessible to DNase I and inactive CHROMATIN is condensed.

DNA sequencing   A method to determine the order of nucleotides on a DNA fragment or molecule. Methods include use of chemicals to break the DNA chains at specific bases (see MAXAM-GILBERT SEQUENCING) and enzymatic incorporation of dideoxynucleotides, which result in chain termination. (See DIDEOXY SEQUENCING.)

docking protein (DP)   A receptor protein, located on the membrane of the ROUGH ENDOPLASMIC RETICULUM (RER), that binds the signal-recognition particle (SRP), a protein-RNA complex bound to the initial sequence of a protein in the process of being synthesized and destined for secretion outside the cell. The docking protein anchors the partially synthesized protein to the membrane of the RER, so as its synthesis is completed it is deposited into the RER where it will be targeted for secretion.

domain   A compact globular unit of protein structure. Many large proteins have domains usually connected by flexible regions of polypeptide chain. The antibody molecule has variable domains that recognize different antigens and constant domains that characterize each class of antibody molecule. (See EPITOPE.)

dominance   The ability of a genetic trait to be phenotypically or physically expressed whether it occurs as HETEROZYGOUS or HOMOZYGOUS. (See RECESSIVE.)

dosage effect The ability of a phenotype to be altered by an increase in the amount of gene product.

dot blot A hybridization technique used to rapidly quantitate the amount of DNA or RNA in a crude preparation placed directly onto a hybridization membrane. (See BLOT.)

double crossover Two recombination events on the same chromosome. (See CROSSING OVER.)

double digestion The treatment of a preparation of DNA with two RESTRICTION ENZYMES. This technique is used to map DNA and to isolate fragments of DNA with two distinct sticky ends, so that it can be cloned into a vector in a particular orientation. (See CLONING, RESTRICTION ENDONUCLEASE.)

double helix Another name for a molecule of DNA, consisting of two antiparallel complementary strands of deoxypolynucleotides held together by hydrogen bonds between the complementary pairs. The molecule has a right-handed twist, resulting in one strand wrapped about the other to form a helix conformation. (See DEOXYRIBONUCLEIC ACID.)

double minutes Small pieces of a chromosome that contain many copies of a particular gene. The amplification of the dihydrofolate reductase (DHFR) gene following exposure to methotrexate may be manifest either in terms of the formation of double minutes or as a HOMOGENEOUSLY STAINING REGION of a giant chromosome.

double reciprocal plot A method of analyzing the kinetic parameters of an enzyme (K_m and V_{max}) by plotting $1/v$ versus $1/[S]$, where v = rate of product formation and [S] = substrate concentration.

double thymidine block A technique used to synchronize cells in culture. A high concentration of thymidine added to the culture will block DNA replication, so all treated cells proceed through their cell cycle and stop at the same point. (See CELL SYNCHRONIZATION.)

doubling time The same as a generation time, or the time it takes for a population of cells to double in number.

down promoter mutation A mutation or change in the sequence of the PROMOTER of a gene, resulting in less expression or transcription of that gene.

downstream Denoting the region of a gene located away from the gene in the direction of the 5'-end.

Drosophila A genus of small flies that includes *Drosophila melanogaster,* the common fruit fly. It is a well-defined genetic organism, used as a model system to study and understand cell processes, development, and genetics of higher organisms.

Drosophila **heat shock proteins (hsp)** Several proteins that are immediately synthesized after the organism is subjected to a short treatment of heat above its lethal limit. Some hsp are highly conserved in evolution, in that they are very similar to hsp found in bacteria and other higher organisms. Synthesis of hsp can also be induced by exposure to certain toxic chemicals, alcohol, and other types of stress. (See HEAT SHOCK PROTEINS.)

drug delivery systems See DELIVERY SYSTEMS.

duplex Another name for double-stranded helical DNA. (See DOUBLE HELIX.)

duplex melting   The process of denaturing double-stranded DNA by heating so that the hydrogen bonds between complementary bases are disrupted. (See DNA DENATURATION.)

dyad   Two units or a pair.

dyad symmetry of DNA   Two regions of the DNA with inverted, repeated, or palindromic base-pair sequences. Restriction enzyme cutting sites exhibit dyad symmetry. (See PALINDROME.)

E

E1A   An adenovirus early gene responsible for the oncogenic properties of the virus.

early development   A stage in the growth cycle of a bacteriophage that precedes DNA synthesis.

early genes   Viral genes that are the first to be expressed after the virus infects its host. They are generally responsible for replication of the virus DNA and for inducing EXPRESSION of the late genes at some specific point in the viral life cycle.

ecdysone   A hormone that induces expression of critical genes during larval development in insects. Known to be responsible for gene transcription in chromosome puffs.

E. coli (Eschericia coli)   A bacterium normally found in the intestinal tract. Because of its ability to grow rapidly under minimal nutritional conditions, its well characterized genetics, and its capacity to host a variety of plasmids and bacteriophages, *E. coli* is widely used as a vehicle for carrying recombinant DNAs and as material for studying bacterial genetics.

ecology   The field dealing with the interrelationship between a population of organisms and the environment, including physical factors and populations of organisms.

EcoRI   A restriction enzyme (see RESTRICTION ENDONUCLEASE) derived from the bacterium *E. coli* with the recognition sequence

GAATTC
CTTAAG

EcoRI methylase   An enzyme that catalyzes the transfer of methyl groups from the compound s-adenosylmethionine to an adenine nucleotide in the restriction site of the enzyme EcoRI:

$$\cdots\text{GAATTC}\cdots \atop \cdots\text{CTTAAG}\cdots \xrightarrow[\text{s-adenosyl-methionine}]{\overset{\text{EcoRI}}{\text{methylase}}} \overset{\text{me}}{\underset{\text{me}}{\mid}} \cdots\text{GAATTC}\cdots \atop \cdots\text{CTTAAG}\cdots \mid$$

ectoplasm   An archaic term for the outer portion of the cytoplasm of a cell.

Edman degradation   A procedure determining the sequence of amino acids in a polypeptide. The procedure is based on reaction of each amino acid in the peptide chain, in order, with the Edman reagent, phenyl isothiocyanate (PITC). The Edman degradation is used in devices for automated polypeptide sequencing.

effector A regulatory molecule. A chemical that brings about an increase or decrease in the rate of reaction in a specific biochemical pathway.

efferent Running in the direction away from a certain structure. For example, efferent nerve fibers carry nerve impulses away from the brain to an effector (e.g., a motor neuron).

effluent Waste fluid such as buffer emerging from a chromatographic column either before or after actual chromatography.

egg The common term for an oocyte.

egg coat A specialized extracellular matrix composed of glycoproteins surrounding the oocyte plasma membrane. In mammalian eggs the egg coat is called the *zona pellucida;* in sea urchins it is called the vitelline layer. In addition to protecting the egg, the egg coat sometimes functions as a selective barrier to fertilization by sperm from different species.

electroblotting A technique that utilizes an electric field to transfer protein or nucleic acids from a gel to a blotting membrane generally for the purpose of carrying out northern, Southern, or western blot HYBRIDIZATIONS.

electrodialysis The technique of accelerating the process of dialysis by applying an electric field across the dialysis membrane.

electrodiffusion The induction of movement of a charged substance by an electric field.

electroendosmosis The diffusion of water into or out of a gel or membrane in the presence of an electric field. Electroendosmosis, resulting in the shrinkage or swelling of an agarose gel, influences

the migration of nucleic acids during agarose gel electrophoresis.

electrolyte A charged atom or molecule in solution.

electron carrier In the biochemical context, a molecule that accepts electrons or hydrogen atoms from a specific donor molecule and then transfers them to a specific electron acceptor. FAD, NAD+ and ubiquinone are important examples.

electronegativity The affinity that an atom or molecule has for electrons.

electronic potential The measure of electron pressure in volts; the relative difference in the concentration of electrons in two compartments, such as the inside of a cell membrane versus the outside of the membrane.

electron microscope A device that utilizes a beam of electrons passing through a specimen, instead of light, to visualize and magnify the features of the specimen. A powerful magnet, used to bend the electron beam, is equivalent to the glass lens in a conventional microscope, which is used to bend the light beam to achieve magnification.

electron microscopy A procedure for using the electron microscope to achieve high levels of magnification. Since electron microscopy must be carried out in a vacuum, biological specimens are generally first coated with a thin layer of metal that conveys the outlines of the structural features of interest.

electron transport The process of passing electrons among electron carriers according to a defined sequence.

electron transport chain A group of electron carriers in the MITOCHONDRION in which the energy released dur-

The Electron Microscope

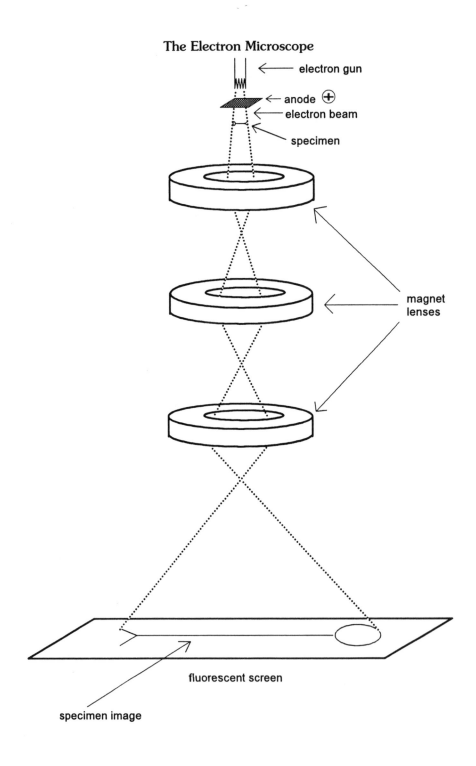

electron gun

anode ⊕

electron beam

specimen

magnet
lenses

fluorescent screen

specimen image

A Elution Profile

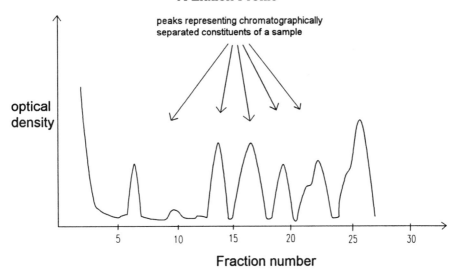

peaks representing chromatographically separated constituents of a sample

optical density

Fraction number

ing the passage of electrons from one carrier to the next is used to create ATP from ADP and inorganic phosphate.

electrophoresis The movement of substances through a medium induced by an electric field.

electroporation A technique for introducing substances into cells by using a pulsed electric field to cause the target substance to be electrophoresed across the cell membrane.

elongation factor Any of several protein factors necessary to carry out that part of TRANSLATION in which amino acids are added to the growing polypeptide chain (elongation).

elongation factors A group composed of at least three proteins (EF-G, EF-Ts, EF-Tu) required for the elongation of a polypeptide in the process of being synthesized on RIBOSOMES (translation).

eluant In column chromatography, the fluid, such as a buffer solution, that

runs through a column and in which separated substances appear as they are washed through the column.

elution profile In column chromatography, a graph showing the amount of material appearing in the eluant of a column over time. The elution profile is generally seen as a series of peaks representing the optical density or biological activity of the eluant at various times during elution of the material undergoing separation.

elution volume In column chromatography, the amount of eluant that passes through a column before a particular peak in the elution profile is observed.

embryo In vertebrates, the organism that develops from the fertilized egg at any stage prior to birth.

embryology The field of study devoted to the development of the embryo.

emulsifier A chemical, such as a detergent, capable of breaking up a mass

Enantiomers (optical isomers)

a molecule containing an atom
with four different substituents

its optical isomer

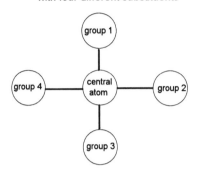

for example:

$$\underset{2}{HN}-\overset{\displaystyle CH_3}{\underset{\displaystyle \underset{HO}{C}\!\!\diagdown\!\!O}{C}}-H \qquad H-\overset{\displaystyle CH_3}{\underset{\displaystyle \underset{HO}{C}\!\!\diagdown\!\!O}{C}}-NH_2$$

of insoluble material into small particles that then form an emulsion. The most common biochemical emulsifiers produce emulsions from otherwise water-insoluble fatty substances.

emulsion For two unmixable liquids, a colloid of one of the liquids suspended in the other (e.g., emulsified oil and water).

enantiomers A pair of isomers that are direct mirror images of one another.

encapsulation The process by which particles become engulfed in or coated by a continuous matrix.

endergonic reaction A chemical reaction that requires the input of en-

ergy, such as heat, or mechanical agitation.

end-filling Creating a blunt end from a ragged-ended or staggered-ended double-stranded DNA through the use of a DNA polymerase.

endocrinology The field of study devoted to the function and pathology of the endocrine glands (e.g., the thyroid and pituitary glands).

endocytic vesicle The membrane, enclosed vesicle that forms in the cytoplasm of a cell during ENDOCYTOSIS.

endocytosis A process in which cells take up small particles or large molecules by an invagination of the cell membrane that leads to the formation of a membrane-enclosed vesicle in the cell cytoplasm. Important effectors influencing cell behavior include the induction of gene transcription after being delivered to the cytoplasm via endocytosis.

endogenous Originating from within; as from within a cell or a tissue.

endogenous virus A virus, usually in latent form, inside a cell. This term applies to various viruses found in an inactive state in the cells they infect but that may become activated following exposure of the infected cells to various chemical and physical agents. This is true of bacteriophage proviruses and of some forms of *Herpesvirus*.

endonuclease A type of enzyme that produces nucleic acid strand breaks in the interior of the nucleic acid strand.

endoplasm The inner part of the cell cytoplasm—that is, the portion closest to the nucleus.

endoplasmic reticulum (ER) A complex cytoplasm membrane network on which ribosomes engaged in the synthesis of proteins destined to be exported outside the cell are attached. Portions of the endoplasmic reticulum containing completed proteins for export are transported to the GOLGI APPARATUS to which they fuse.

end product The final chemical product of the series of enzymatic reactions in a particular biochemical pathway.

end-product repression See FEED-BACK INHIBITION.

endorphin Any of a group of short peptides that bind to receptors on neurons in the brain with the effect of reducing the sensation of pain. The term is derived from "endogenous morphine." (See METHIONINE-ENKEPHALIN.)

endosome The structure formed by the fusion of several endocytic vesicles in the cytoplasm following endocytosis.

(See CLATHRIN, COATED PIT, COATED VESICLE.)

endospore A tough, resistant, membrane-enclosed cell formed by some gram-positive bacteria and actinomycetes under conditions of limited food supply. The endospore is highly dehydrated and metabolically inactive and can survive harsh environmental conditions, such as prolonged heat, drying, and exposure to toxic chemicals.

endosymbiont A symbiotic organism living inside the body of its symbiotic partner.

endothermic Describing a chemical reaction that requires heat in order to proceed.

endotoxins A group of lipopolysaccharides in the outer membranes of gram-negative bacteria. The main pathological effects of bacterial endotoxins are diarrhea and hemorrhagic shock.

enhancers Certain DNA nucleotide base sequences that act over distances up to several kilobases (1000 bases) to stimulate transcriptional activity of a gene or group of genes.

enriched medium A supplemented nutrient broth for the culture of cells or microorganisms that require unusual nutrients or unusually high levels of normal nutrients; required for the culture of auxotrophic mutants. (See AUXOTROPH.)

enteric organism A microorganism inhabiting the intestinal tract.

Enterobacteriaceae Any of a large group of bacteria inhabiting the intestinal tract.

entomology The science dealing with insects.

entropy The variable that measures the degree of disorder in a molecule or in a system. Changes in entropy occurring in molecules undergoing chemical reaction are one component of the free-energy change that determines whether a reaction will occur under a given set of conditions.

env gene(s) One of three genes in most retroviruses that codes for the env GLYCOPROTEINS. (See RETROVIRUS.)

env glycoproteins The protein product of the retrovirus ENV gene(s) that forms a major component of the virus envelope in the mature virus particle.

enzyme A polypeptide or protein that catalyzes (speeds up) biochemical reactions. Virtually all significant biochemical reactions in living systems are catalyzed by enzymes.

enzyme derepression The induction of enzyme activity by removing or inactivating an inhibitor, such as the induction of β-galactosidase activity by lactose. (See LAC OPERON.)

enzyme immobilization The chemical bonding of an enzyme to some solid matrix in a manner that preserves enzymatic activity. Attaching enzymes to solid matrices is an essential step in developing many enzyme-based biochemical assays.

enzyme inactivation Loss of enzyme activity under conditions other than those found in the intact cell. Enzyme inactivation is an important consideration when purified enzymes are employed in an environment where they may be subject to conditions of temperature, salt, pH, and so on, not found in their native environment. Spontaneous inactivation of enzymes occurring for unknown reasons is also often observed in enzyme preparations, particularly in dilute solutions.

enzyme linked immunoabsorbent assay (ELISA) A sensitive technique for detecting a substance by allowing it, if present in a sample, to attach to an immobilized antibody on some solid substrate, such as plastic. The presence of the substance is visualized and quantitated by using a second, labeled, antibody.

enzyme replacement therapy The method used to treat disease states caused by enzyme deficiencies by direct injection of the missing enzyme. Enzyme replacement therapy has been used successfully for treating patients with Gaucher's disease.

enzyme stabilization Inhibition of enzyme inactivation. Enzyme stabilization is often achieved by altering the salt concentration, the pH, or lowering the temperature of an enzyme solution. Recently, modification of the enzyme by attachment of organic groups or altering the amino acid composition of the enzyme polypeptide have been used to achieve enzyme stabilization.

eosinophil One of three subclasses of leukocytes, named for its characteristically intense staining with eosin. Eosinophils are amoeboid scavenger cells similar to macrophages and are found in greatly increased numbers in the blood of individuals carrying parasitic infections.

epidemiology The science devoted to analysis of the occurrence of disease in a population, including the distribution, incidence, and factors controlling spread of disease.

epidermal growth factor (EGF) A small polypeptide growth factor, discovered by Stanley Cohen, causing premature eyelid opening in newborn mice. EGF has since been shown to be active in stimulating the growth of epithelial as well as some nonepithelial cell types. A portion of the gene that codes for the EGF cell receptor has been found to be virtually identical to the *erb-B* oncogene.

epigenetic The term applied to any factor that influences cell behavior by means other than via a direct effect on the genetic machinery (i.e., the DNA).

epimerase A type of enzyme that catalyzes the conversion of one epimer into its opposite epimer.

epimers Optical isomers that differ from one another at only a single carbon

Enzyme-linked Immunoabsorbent Assay (ELISA)

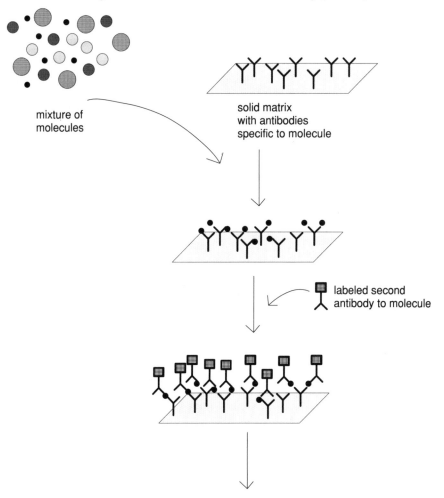

mixture of
molecules

solid matrix
with antibodies
specific to molecule

labeled second
antibody to molecule

quantitate amount of labeled antibody bound

atom. The sugars glucose and galactose are examples of epimers.

epinephrine (adrenaline) Biochemical secreted by the adrenal glands and by the synaptic vesicles of certain types of neurons. Epinephrine serves as both a hormone stimulating the breakdown of glycogen into glucose and as a neurotransmitter.

episome Bacterial DNA that is not integrated into the bulk of the chromosomal DNA and therefore replicates sep-

arately, and in different copy number, from chromosomal DNA.

epistatic gene A gene that suppresses the effect of another, nonallelic, gene. (See ALLELE.)

epithelial Of, or pertaining to, the cell layers that interface between the tissue and the external environment, such as the cells of the skin, the lining of the gut, and lung airway passages.

epitope The segment on a polypeptide that constitutes the actual site of

Epinephrine (adrenaline)

antibody binding by a specific antibody molecule; the antigenic determinant.

Epstein–Barr virus   A member of the Herpes family of DNA viruses that has been associated with Burkitt's lymphoma in West Africa and New Guinea.

equatorial plate   The early stage of the formation of the membrane that divides two daughter cells at the end of MITOSIS; the metaphase plate.

equilibrium centrifugation   A technique for separation of proteins or nucleic acids from a mixture by subjecting the mixture of density gradient centrifugation for a sufficient time for each component of the mixture to form a band at a point equal to its density.

equilibrium potential   The membrane potential at which there is no net diffusion of a particular type of ion across the membrane. Equilibrium potentials are important determinants of nerve impulse generation.

erb-a   One of the two oncogenes carried by the avian erythroblastosis RETRO-VIRUS. The *erb-a* oncogene has been found to be virtually identical to the gene that codes for the thyroid hormone receptor protein.

erb-b   One of the two oncogenes carried by the avian erythroblastosis RETRO-VIRUS. The *erb-b* oncogene has been found to be virtually identical to the gene that codes of the epidermal growth factor receptor protein.

error-prone repair   Another term for SOS repair. The terminology is derived from the observation that repair of pyrimidine dimer damage is often inac-

curate. (See EXCISION REPAIR, SOS REPAIR SYSTEM.)

erythroblast   A bone marrow stem cell giving rise to erythrocytes.

erythrocyte   A red blood cell.

erythrocyte ghosts   Red blood cells whose contents have been removed. Erythrocyte ghosts are used as vehicles to deliver drugs and other bioactive compounds to cells. (See DELIVERY SYSTEM.)

erythromycin   An antibiotic that acts by binding to bacterial ribosomes and inhibiting the process of translocation during protein synthesis.

erythropoesis   The process by which erythrocytes are generated from stem cells in the bone marrow.

erythropoetin   A glycoprotein, produced by the kidney, that stimulates erythropoesis.

Escherichia coli   See E. COLI.

essential amino acid   Any amino acid that cannot be synthesized by an organism from other components. In humans, about 10 of the 20 amino acids are essential, whereas in most bacteria none are.

essential gene   Any gene whose malfunction is lethal to an organism. A number of classical experiments on bacterial molecular genetics such as fluctuation analysis depended on the use of mutations in essential genes.

established cell line   Cells that have become immortalized during the process of maintaining them in cell culture.

establishment of cell lines   The process by which cells in tissue culture become immortalized so that they can be maintained indefinitely. Establishment is believed to involve some genetic change that occurs spontaneously during the course of culture. Because cells derived from cancerous tissue are more readily established than cells from normal tissue, the genetic changes involved in the process of becoming established are also believed to be related to the process by which cells become cancerous.

esterase   A type of enzyme that catalyzes the breakage of ester linkages. Esterases are important in the breakdown of many lipids and in the metabolism of nucleic acids.

estrogen   A steroid hormone produced by the ovaries, causing changes in the lining of the uterus in preparation for implantation of the embryo during estrus.

estrus cycle   A set of changes occurring periodically in female primates that prepares the reproductive tract for pregnancy and is governed by changes in the levels of female hormones. Cycle peak (called estrus) coincides with ovulation.

ethanol   Ethyl alcohol (drinking alcohol); the alcohol produced during the fermentation of sugar by certain strains of anaerobic yeast.

ethidium bromide   A widely used fluorescent stain for visualizing DNA

Intercalation of Ethidium into DNA

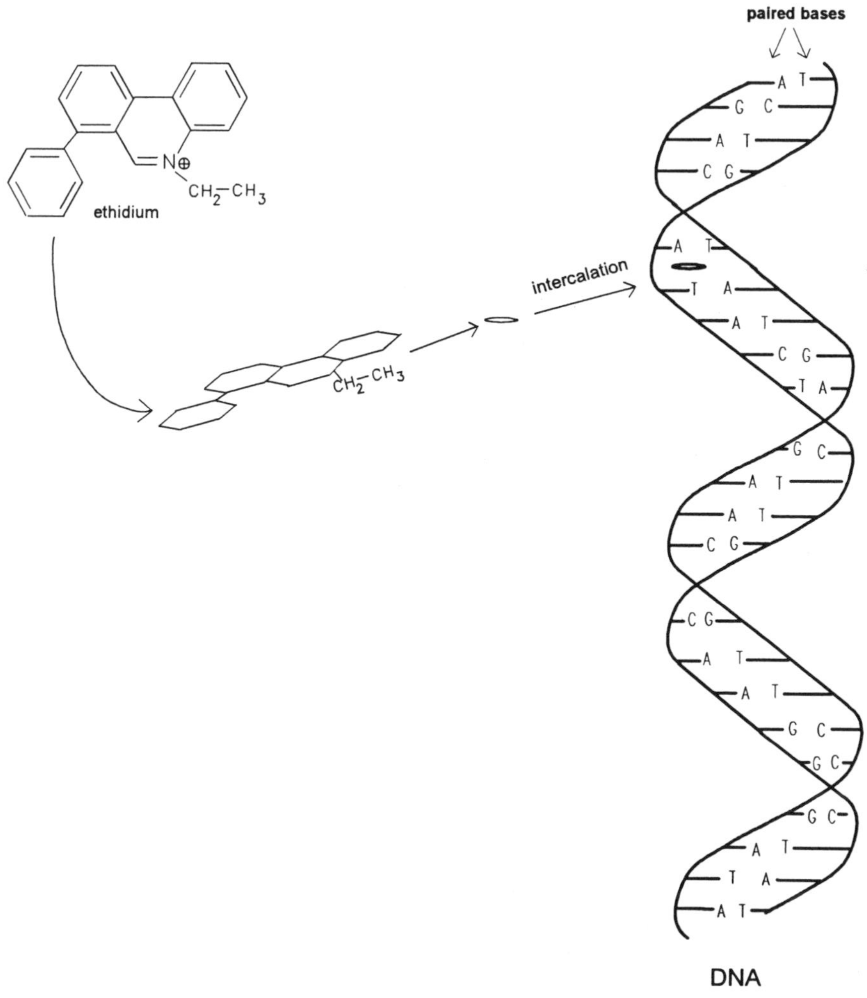

DNA

under ultraviolet light. Ethidium bromide is called an intercalating dye because it has a multiring structure that allows it to insert between the nucleotide bases.

ethylene   A simple two-carbon hydrocarbon with formula $HC\!\!=\!\!CH$

ethylene diamine tetra acetate (EDTA)   A chemical that binds tightly to magnesium and calcium and is used to effectively remove even trace amounts of these metals from a solution. EDTA is used to control unwanted magnesium- and calcium-dependent side reactions in a biochemical mixture.

etiology   The study of the cause of a disease or pathological condition.

ets oncogene   An oncogene carried by avian leukemia virus E26 (v-ets), which causes leukemias in chickens. The product of the ets proto-oncogene (c-ets) is a nuclear protein found to have DNA binding activity and believed to play a role in the activation and proliferation of T cells.

euchromatin   One of two chromatin classes seen in interphase cells and distinguished from the other class (HETEROCHROMATIN) by being much less condensed and transcriptionally active.

eugenics   The science of selective breeding to achieve a predetermined set of genetic characteristics.

euglena   A primitive single-celled organism classified as belonging to the algae in the plant kingdom. Euglena exhibits the properties attributed to both plants and animals; that is, it is photosynthetic in the presence of light and a motile, food-seeking organism in the absence of light. Euglena is believed to represent, or be related to, ancestral organisms that gave rise to both plants and animals.

eukaryote   The general term for any higher plant or animal distinguished by the presence of a true nucleus containing the DNA. Bacteria and viruses are the organisms that comprise the noneukaryotes (i.e., prokaryotes).

eukaryotic cell   any cell in which the cellular genome is contained in a membrane-enclosed nucleus. In general, eukaryotic cells include all cells in the plant and animal kingdoms other than bacteria and viruses.

European Molecular Biology Lab (EMBL) gene bank   A large DNA sequence database containing sequence data compiled from international sources and maintained in Heidelberg, Germany; the European equivalent of the Genbank DNA sequence databank.

evolution   In the biological context, the term evolution is generally equated with natural selection as proposed by Darwin: the process of change in the composition of a population resulting from the selection of a subpopulation, better fit than the population as a whole, for survival under a particular set of environmental conditions.

excision repair   The process of repairing damaged regions of DNA, which involves excising the damaged region, recopying the excised region by DNA polymerase and ligating the recopied region by DNA ligase. (See ULTRAVIOLET REPAIR.)

exergonic   A chemical reaction that releases energy in various forms, such as heat or light.

exit domain   One of the two major classes of binding sites on the RIBOSOME. The finished polypeptide, which is the product of the process of TRANSLATION, leaves the ribosome at this site.

exocellular   Pertaining to processes or reactions that originate within the cell but take place outside the cell. For example, the digestion of EXTRACELLULAR proteins by proteolytic enzymes secreted by a cell is an exocellular event.

exocrine   Secretion of a glandular-produced substance via a duct or canal

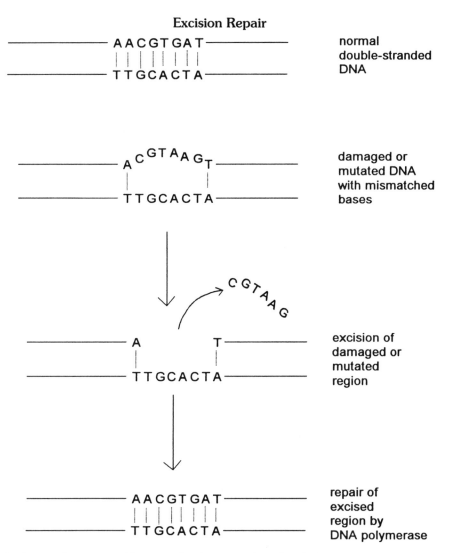

Excision Repair

leading to the exterior. Exocrine glands are distinguished from endocrine glands, which secrete their products into the bloodstream. Sweat produced by eccrine glands and milk produced by mammary glands are examples of exocrine secretion.

exocytosis The process in which substances contained in a specialized vesicle within the cytoplasm of a cell are secreted to the outside by fusion of the vesicle with cell plasma membrane. The secretion of neurotransmitters in synaptic vesicles by neurons is a common example of exocytosis.

exon The regions of a gene in eukaryotic cells that, as opposed to INTRONS, contain the coding sequences for a polypeptide. (See SPLICING.)

exonuclease A class of enzymes that catalyze the cleavage of nucleotides from the end(s) of a nucleic acid.

λ exonuclease An enzyme that catalyzes the cleavage of single nucleotides with 5'-phosphate groups from the 5'-ends of double-stranded DNA.

exonuclease III An exonuclease that catalyzes the cleavage of single nucleotides one at a time from the 5'-end of double-stranded DNA, which has a non-phosphorylated 3'-end.

exonuclease VII An exonuclease that catalyzes the cleavage of short oligonucleotides from the 3'- and 5'-ends of single-stranded DNA.

exoskeleton A general term referring to a solid, inert matrix that completely surrounds the soft tissue of an organism and therefore forms a shell, for purposes of tissue support and protection. The shells of crabs, lobsters, and other crustaceans are examples of exoskeletons.

exotoxin Any of a variety of toxic substances produced by a microorganism and released to the surrounding fluid.

explant The growth of a portion of a tissue away from its normal location.

exponential growth phase The GROWTH PHASE of a population of cells or organisms during which the overall population number is seen to double at a regular interval. (See BIPHASIC GROWTH.)

export The transport of substances across the cell membrane from the interior of a cell to the exterior via specialized systems.

expression See GENE EXPRESSION.

expression library A library of DNA fragments that has been created by using a vector designed to express any genes that are present in the library. (See EXPRESSION SYSTEM, EXPRESSION VECTOR.)

expression-linked copy (ELC) The particular VARIABLE-SURFACE GLYCOPROTEIN gene being expressed at any time during the developmental cycle of the TRYPANOSOME.

expression site The term for the genetic location of an EXPRESSION-LINKED COPY of a variable surface glycoprotein. Expression sites are all located near the TELOMERE of a chromosome.

expression system An EXPRESSION VECTOR, containing the cloned DNA it is designed to express, together with the host with which the vector is to be used.

expression vector A specialized cloning vector containing the elements needed to transcribe a cloned DNA. Expression vectors contain sequences required for DNA replication and promoter elements adjacent to the cloned DNA in order to initiate transcription.

extinction coefficient The constant of proportionality relating the molar concentration of a substance and the absorbance of its solution. (See BEER–LAMBERT LAW.)

extracellular Outside the cell.

extracellular fluid The liquid outside the cells in a tissue.

extracellular matrix A complex mixture of proteins (such as FIBRONECTIN, LAMININ, collagen) deposited on the outside of a cell that plays a crucial role in the attachments of cells to the surfaces on which they grow. The extracellular matrix is believed to play an important role in regulating the growth and differentiation of a cell partly because the composition of the extracellular matrix is often dramatically altered in cancerous tissue.

extrinsic protein A protein present in a cell or tissue but that originated elsewhere.

extrusion The energy-requiring process by which cells export large particles or organelles.

F

F(ab)₂ fragment A portion of the IgG antibody molecule containing the two antigen-binding domains but not the Fc portion. Such fragments are generally produced by treating antibodies with certain proteases that specifically cleave the molecule near the end of the Fc segment.

facilitated diffusion The process of passive diffusion through a membrane via membrane channels or with the aid of carrier proteins in the membrane.

f-actin (filamentous actin) The functional actin filament composed of G-actin subunits.

factor, blood clotting Any of a group of protein factors in the blood serum that act according to a defined pathway to produce a blood clot. In blood clotting the breakdown of platelets at the wound site is the first step. Factors VII, VIII, IX, and XI become activated by tissue factor and, in the presence of calcium, convert factor X to activated thromboplastin. Thromboplastin then converts prothrombin into thrombin, the clotting enzyme. Thrombin catalyzes the conversion of soluble fibrinogen in the serum into the insoluble protein fibrin. Fibers made of fibrin are the basic structure of the final clot and are made firm by factor XIII, or fibrin-stabilizing factor. The genetic disease hemophilia is the result of a defect in the gene that codes for factor VIII.

facultative anaerobe A microbe living under anaerobic conditions that can adjust its metabolism to utilize oxygen when placed in an aerobic environment.

facultative heterochromatin A highly condensed form of heterochromatin believed to not be transcribed.

facultative microorganisms See AEROBE.

fast component That portion of a preparation of eukaryotic DNA that reassociates first when the strands of a native DNA helix are dissociated from one another (e.g., when heated) and then allowed to reassociate. The fast component was shown to contain highly repeated DNA sequences.

fastidious Pertaining to microorganisms with complex nutritional requirements; requiring enriched media. (See ENRICHED MEDIUM.)

fatty acid Any of a group of long-chain carboxylic acids found mostly in animal fat; the most common are oleic, palmitic, stearic, and palmitoleic acid.

Fc The portion of the immunoglobulin heavy-chain molecule that does not contain the antigen-binding region; the antibody molecule constant region.

fecundity the measure of fertility—sperm count or the production of viable eggs.

feedback control The general term for regulation of an enzyme's activity by one of its own metabolites.

feedback inhibition Inhibition of enzyme activity by a product (generally the final product) of the metabolic pathway that the enzyme is part of.

feeder layer In tissue culture, a layer of cells that produces a product supporting the growth of another cell type in coculture. Feeder layers are used as a means of growing cells that will not grow in purely synthetic culture medium.

feedforward control Stimulation of enzyme activity by a product of the meta-

Feedback Inhibition

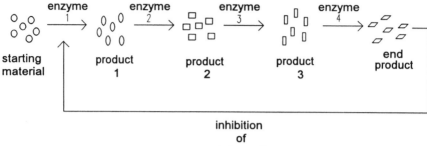

bolic pathway that the enzyme is part of.

feline sarcoma (fes)   The oncogene carried by Snyder–Theilen strain of feline sarcoma virus (FSV). It is believed to be a PROTEIN KINASE that catalyzes the phosphorylation of tyrosine residues.

feline sarcoma virus (FSV)   retrorus that causes sarcoma tumors in cats.

fermentation   Microorganic process in which the metabolism of sugars for energy is accompanied by the formation of alcohol or lactic acid.

fermentor   A device for growing large bacterial cultures. It consists of a large vessel (usually containing more than 10 liters of culture) that is mechanically shaken or rapidly rotated for aeration of the culture. There is also a heater in contact with the culture vessel that maintains the culture at the proper temperature, usually 37°C.

ferritin   A protein that forms a complex with iron. It normally functions as an iron storage protein and is used as a nonradioactive label for visualizing antibodies bound to a specific antigen, such as in a western blot.

fertilization   The fusion of two gametes and their respective nuclei to create a diploid or polyploid ZYGOTE.

fetal calf serum (FCS)   The serum from the blood of embryonic calves, an essential component of most tissue culture media. The factors in FCS that promote the growth of cells in tissue culture are largely unknown but are believed to include growth factors and hormones.

Feulgen reagent   A DNA-specific stain (fuchsin sulfite) that, because it was found to stain CHROMATIN in the nucleus strongly, was cited as evidence that DNA was the hereditary material in experiments carried out by Robert Feulgen in 1914.

F-1 generation   First filial generation; the first-generation offspring of a genetic cross. The term is generally applied to the immediate offspring of higher plants and mammals.

F-2 generation   Second filial generation; the offspring of a mating between members of the F-1 generation.

F-factor   A small episomal segment of DNA that functions as a bacterial mating factor. The F-factor functions by inserting itself into the bacterial chromosome, which can then be transferred to an adjacent bacterium. (See HIGH-FREQUENCY RECOMBINATION STRAIN.)

fgr   The oncogene carried by Garden–Rasheed strain of the feline sarcoma virus.

fibrin The protein formed from fibrinogen that polymerizes to form the fibers that compose a blood clot at the site of a wound.

fibrinogen The protein released by platelets at the site of a wound and giving rise to fibrin when thrombin is present.

fibroblast A cell type that composes the bulk of the living cells in connective tissue and in the supporting matrix (the stroma) of skin and other epithelial tissues. Fibroblasts are embedded in a complex extracellular matrix, much of which they secrete, responsible for the strength and flexibility of the stroma.

fibronectin A ubiquitous glycoprotein found in blood and in virtually all tissues of the body. It is thought to play a key role in cell adhesion and in the control of cell growth and differentiation. Fibronectin is particularly prominent in fibroblast-containing tissues where it is complexed with collagen.

Fick's law of diffusion The premise that a substance in solution will diffuse in a direction that will tend to eliminate any concentration gradient—that is, make the solution homogeneous with respect to concentration.

ficoll A synthetic polymer of the sugar, sucrose, it is biochemically inert and used primarily to increase the density of solutions for density gradient centrifugation and nucleic acid hybridization.

ficoll gradient A solution of ficoll created in such a way that the concentration of ficoll varies continuously along an axis through the solution. Ficoll gradients are often used to separate different cell types from one another by sedimentation. (See DENSITY GRADIENT.)

figure eight An intermediate stage in the process of recombination in which two circular DNAs are covalently bound to one another.

filamentous bacteriophage A subclass of single-stranded DNA BACTERIOPHAGE in which the bacteriophage genome is encapsidated by an elongated viral coat resembling a filament. F1 and M13 are the most common members of this class of bacteriophage.

filopodia Long microspikes (50 μm) extending out of the growing tip of the axon of a developing neuron.

filter sterilize A technique for rendering a solution sterile (i.e., free of microbes the size of bacteria) by passing it through a fine filter.

fine structure mapping A mapping technique that can detect changes in nucleotide base sequence covering a few nucleotides based on very rare recombination events between strands of DNA carrying different forms of the same gene (alleles).

fingerprinting A general term for techniques that define a unique identity for a given protein or nucleic acid molecule by breaking the molecule into a pattern of fragments based on its amino acid or nucleotide base sequence using various proteases or restriction enzymes. Fingerprinting has been developed as a tool in forensic medicine primarily in the form of DNA fingerprinting in which an individual's unique pattern of DNA fragments is visualized by SOUTHERN BLOT HYBRIDIZATION using a probe for a gene that is known to vary widely.

finger protein A protein containing segments of regularly spaced cysteine amino acids that appear to be involved in binding zinc atoms. This type of structure is characteristic of nucleic acid binding proteins. (See ZINC FINGER.)

first-order kinetics Any chemical reaction in which the rate at which the reaction occurs is proportional to the molar concentration of only one reactant. For example, for the reaction A → B + C, the rate of reaction = $K[A]$, where k is the reaction rate constant.

Flavin Adenine Dinucleotide (FAD)

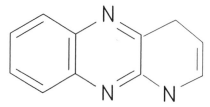

$$CH_2-O-\overset{\overset{O}{\|}}{\underset{\underset{OH}{|}}{P}}-O-\overset{\overset{O}{\|}}{\underset{\underset{OH}{|}}{P}}-O-CH_2$$

flagella   Long, external, flexible filaments used to propel cells in a liquid medium. Bacterial flagella, which differ in structure from the flagella in eukaryotic cells, also serve as a chemotactic organ that guides the cell to sources of food.

flagellin   The protein constituting the bacterial flagellum.

flash evaporator   A device for removing solvent from large volumes of a solution by evaporation in order to concentrate the solute. A flash evaporator consists of a heated, rotating glass sphere with a tube to allow the evaporating solvent to exhaust.

flavin   A compound derived from riboflavin (vitamin B$_2$). Important flavin biomolecules are FAD and FMN.

Flavin Molecule

flavin adenine dinucleotide (FAD, FADH$_2$)   A cofactor for enzymes involved in oxidoreductions and electron transfer in numerous biochemical reactions, but particularly those involved in the oxidative metabolism of sugars for energy production. FAD is a combination of two nucleotides, one of which is derived from the B vitamin, riboflavin (vitamin B$_2$).

Flavin mononucleotide (FMN)   A nucleotide derived from the vitamin riboflavin (vitamin B$_2$), it is one of the two nucleotides of FAD that also functions in the transport of electrons during the oxidative metabolism of sugars for energy production.

flocculation   The rapid precipitation out of solution of large amounts of material.

flora   Plant life.

flow cytoenzymology   A technique for separation and analysis of cells by fluorescence-activated cell sorting (FACS) based on the presence of certain enzymes generating colored compounds from synthetic substrates.

flow cytometry   A technique based on automated measurement of fluorescence emitted by individual cells. Flow cytometry is carried out in an instrument in which individual cells are illuminated by a laser beam as they pass by a window where a sensitive photocell records the quantity of light emitted at a given wavelength. Because antibodies can be labeled with fluorescent compounds, this technique has been widely used as an automated procedure for quantitating amounts of various antigens present in a population of cells.

fluctuation analysis A method, developed by Salvatore Luria and Max Delbruck in 1943, using statistical analysis of the rate of mutation occurring in bacterial cultures containing small numbers of cells to demonstrate that mutations occur spontaneously.

fluid mosaic model A model of the eukaryotic cell membrane proposed by S. J. Singer and G. L. Nicolson in 1972. The model is based on the idea of a semisolid lipid bilayer into which transmembrane and integral membrane proteins are embedded.

fluorescence The property of certain molecules whereby they emit light at a specific wavelength (emission wavelength) when illuminated by a light beam at another specific wavelength (excitation wavelength).

fluorescence-activated cell sorting (FACS) A variation of flow cytometry in which cells with a fluorescent label are physically sorted into different compartments based on the amount of fluorescence emitted at a given wavelength.

fluorescence recovery after photobleaching (FRAP) A technique whereby fluorescent molecules located in a specific cellular structure (e.g., the nucleus or cell membrane) are bleached by a microscopic light beam. The bleached area is examined at various periods of time after the photobleaching to determine how fast the cellular structure regenerates the material in the

bleached area. The FRAP technique is best known for its use in studies of cell membrane synthesis and fluidity.

fluorescent antibody techniques Techniques for visualization of the location of certain antigens in a tissue section or other cell preparation based on the binding of an antibody with a fluorescent label to the antigen of interest.

fluorescent label Any molecule that fluoresces and can be attached to another, nonfluorescing probe molecule, such as an antibody or DNA hybridization probe.

fluorimetry Quantitative measurement of fluorescence.

5-fluorodeoxyuridine The nucleotide derivative of 5-fluorouracil formed in cells treated with 5-fluorouracil. 5-Fluorodeoxyuridine and 5-fluorouracil are both used as anticancer agents.

5-fluorouracil (5-FU) An analog of thymine used as an anticancer agent. 5-FU is an inhibitor of the enzyme dihydrofolate reductase (DHFR) and, therefore, is an inhibitor of nucleotide synthesis and particularly harmful to the rapidly growing cells in tumors.

flush ends Termini of a double-stranded DNA molecule that have no single-stranded overhanging regions. (See BLUNT-END DNA.)

fms The oncogene carried by the McDonough strain of feline sarcoma virus.

focus-forming assay A test for the presence of DNA containing oncogenic activity. In a focus-forming assay, test DNA is transfected into animal cells that ordinarily show contact inhibition. If the test DNA contains oncogenic activity, the recipient cells lose CONTACT INHIBITION, begin to divide, and then form areas of dense packing (foci). The appearance of foci is taken as an indication of oncogenic activity. (See TRANSFEC-TION.)

focus-forming units (FFU) A measure of the concentration of live virus in a given volume of fluid. Focus-forming units are determined by spreading a known amount of virus-containing fluid over a layer of cells infected by the virus and then observing the number of areas in the cell layer that show evidence of viral infection. (See TITER.)

folate antagonist A type of compound (e.g., methotrexate) that blocks certain critical reactions in the synthesis of nucleotides requiring the B vitamin folic acid. Folate antagonists are widely used as chemotherapeutic agents for treatment of cancer because the rapidly growing cells of malignant tumors depend more on these reactions than do normal cells.

foldback DNA A segment of DNA containing palindromic repeat sequences that may base-pair to one another during reassociation. (See PALINDROME.)

follicle cells A layer of cells found in vertebrates and invertebrates that surrounds the oocyte and supplies it with certain low molecular weight nutrients.

footprint The region on a DNA molecule to which a particular regulatory protein binds.

form(s) I, II, and III DNA For circular DNAs extracted from viruses and plasmids: form I is supercoiled native DNA; form II is circular, nicked DNA; form III is linear DNA produced as an artifact of the extraction procedure caused by breakage of both DNA strands.

formamide (HCONH$_2$) One of the most commonly used chemicals for denaturing nucleic acids in hybridization techniques.

forming face The side of the Golgi stack where vesicles that have budded off from the rough ENDOPLASMIC RETICULUM fuse to the GOLGI APPARATUS; the cis face of the Golgi apparatus.

formycin B A purine derivative used as an antiparasitic agent. It inhibits the ability of cells to use salvaged nucleotides from the extracellular medium for nucleic acid synthesis.

forward mutations Any mutation that causes a change from a normal functioning gene to an improperly functioning, or inactive, gene.

fos The oncogene carried by FBJ murine osteosarcoma retrovirus. The *fos* oncogene product appears to function in concert with the *jun* oncogene to alter rates of transcription of certain genes.

fosfomycin (fosfonomycin) An antibiotic that acts by blocking an early step in the biochemical pathway by which the bacterial cell wall is synthesized. Fosfomycin, a structural analog of phosphoenol pyruvate (PEP), blocks the step at which PEP is required to create the pentapeptide used to construct the bacterial cell wall.

fos-related antigens (FRA) A group of nuclear phosphoproteins similar in structure to the product of the *fos* oncogene.

founder effect The presence of a chromosome, portion of a chromosome, or even a particular allele in the members of a given population that can be traced back to a single individual.

four-strand crossing over CROSSING OVER between two sister chromatids that involves breaking of both DNA strands on both chromosomes. This differs from the usual case involving only one DNA strand from each CHROMATID.

F protein The SENDAI VIRUS–derived protein responsible for the ability of the virus to cause cell fusion, and used in the creation of FUSOGENIC VESICLES.

fps The oncogene carried by the Fujinami sarcoma virus. The *fps* and *fes* oncogenes are homologous genes from chicken and cat, respectively; the proteins coded for by these genes are tyrosine kinases.

fragile sites Sites on chromosomes that show a higher than normal probability of breakage and therefore more commonly sites where chromosomal translocation is observed.

frameshift mutation A type of mutation in which nucleotide bases are inserted or deleted in the coding region of a gene, causing the triplet CODONS to be translated in the wrong reading frame.

free energy See GIBBS FREE ENERGY.

freeze drying The removal of ice from a frozen cell sample to be examined by the freeze-etch technique by subjecting the sample to a vacuum as the temperature is slowly raised, thereby leaving the essential cell structural features behind. (See LYOPHILIZATION.)

freeze etch A technique for examining cell structure by electron microscopy in which a frozen cell sample is cracked with a knife to reveal the cell contents. After freeze drying, the sample is then shadowed and examined under the electron microscope.

freeze fracture A technique for examining the structure of the cell membrane by electron microscopy. The procedure is essentially the same as in freeze etching except that the sample is fractured along the plane of the cell membrane and is then examined after freeze drying and shadowing.

Frei test A clinical test to diagnose diseases caused by infectious microbes based on the appearance of a skin reaction when a killed preparation of the suspect microorganism is injected subcutaneously.

French pressure cell A device for lysing bacterial cells by subjecting them to hydrostatic pressure. (See LYSE.)

Freund's adjuvant An emulsion consisting of water, oil, and dead mycobacteria that, when mixed with an immunogen, enhances the immune response when the IMMUNOGEN–Freund adjuvant mixture is injected into an animal.

fructose "Fruit sugar." An isomer of glucose found in citrus fruits. A phosphorylated form of fructose is an intermediate metabolite in the oxidation of glucose for energy production.

fumarase The enzyme that catalyzes the conversion of fumarate to malate, an important step in the Krebs cycle phase of sugar metabolism.

fungi See MOLD.

fungicide An agent that selectively kills fungi.

furanose A ring form of a sugar in which the ring is made up of four carbon atoms and one oxygen. The term designates a large group of sugars that form this type of ring when dissolved in water.

fushi tarazu (ftz) gene A gene in the pair-rule locus of the fruit fly, *Drosophila melanogaster*. Mutants of the *ftz* gene are missing every other segment. (See PAIR-RULE MUTANTS, SEGMENTATION, SEGMENTS.)

fusidic acid An antibiotic that blocks the translocation step in protein synthesis (translation) by blocking the release of the elongation factor (EF)–GDP complex.

fusion proteins Proteins representing the product of the artificial splicing of two genes.

fusogenic vesicles Liposomes that contain, in the lipid bilayer, specialized fusion-inducing molecules (e.g., the F protein).

G

galactose An optical isomer of the sugar glucose. Galactose differs from glucose only at the fifth carbon and is converted to glucose through the action of an epimerase enzyme (UDP-glucose-4-epimerase) that acts on this carbon.

galactosemia A genetic disease caused by a deficiency of the epimerase enzyme that converts galactose into glucose. The disease is characterized by organ enlargement, mental retardation, and cataract formation resulting from the accumulation of D-galactose and D-galactose-1-phosphate in the bloodstream.

galactosidase (gal) Any of the class of enzymes that catalyze the cleavage of the glycosidic linkage between galactose and another sugar. The galactosidases are divided into α- and β-galactosidases, depending on the type of glycosidic bond cleaved (i.e., α or β). (See GLYCOSIDIC LINKAGE.)

β galactosidase An enzyme that catalyzes the hydrolysis of the disaccharide lactose to the monosaccharides glucose and galactose. The gene encoding this enzyme in *E. coli* is part of the LAC OPERON.

gamete The mature product of the process of meiosis (e.g., egg and sperm) in organisms that reproduce sexually.

ganglion A collection of neurons that, in mammals, are centers of lower brain function outside the brain proper. In lower animals lacking a brain, such as worms and invertebrates, ganglia constitute the centers of all brain function.

gangliosides A class of cell membrane lipids found almost entirely in brain neurons. Ganglioside molecules form part of the brain receptor complex for pituitary polypeptide hormones.

gap genes A group of genes (hunchback, krüppel, knirps) that play a key role in the development of segmentation in the embryo of the fruit fly, *Drosophila melanogaster*. Gap genes have been identified on the basis of mutations that result in the absence of segments in the midportion of the embryo.

gap junction A specialized channel that forms between two adjacent cells at their mutual point of contact and connects the cells so that small molecules (about 2000 kilodaltons or less) can pass between them. Gap junctions are believed to be a mechanism of intercellular communication involved in the control of cell growth and differentiation.

GAP mutants Mutants of the fruit fly, *Drosophila melanogaster,* in which several adjacent segments fail to appear during the course of development. See SEGMENTATION, SEGMENTS.)

gas chromatography (GC) Another term for gas-liquid chromatography (GLC).

gas-liquid chromatography (GLC) A type of chromatography in which substances in a sample are evaporated into a stream of an inert gas, such as argon, helium, or nitrogen, and then separated at a high temperature by passing the evaporated material through a column containing a liquid, such as silicone oil or polyethylene glycol, on an inert matrix material, such as fire brick. (See CHROMATOGRAPHIC TECHNIQUES.)

gasohol A mixture of 90% gasoline and 10% ethyl alcohol. Gasohol is purported to be a cleaner, more energy efficient alternative to gasoline. It has been proposed that ethyl alcohol for this purpose could be cheaply obtained by bacterial fermentation.

GC box A sequence of nucleotides (GGGCGG) in the PROMOTERS of mammalian cells and that appears to be a binding site for certain TRANSCRIPTION factors.

GC content   The fraction of the total nucleotides in a DNA molecule that are either cytosine or guanine nucleotides (generally given as a percentage of the total).

gel   A semisolid colloidal solution.

gel electrophoresis   A technique for separating substances, principally nucleic acids and proteins in a mixture, by using an electric field to induce them to migrate through a gel. Separation of the individual component substances in the original mixture is based on the size of the molecules (gel filtration) and/or electric charge. Agarose gels are commonly used for separation of mixtures of nucleic acids, and polyacrylamide gels are generally used for separation of proteins and nucleic acids.

gel exclusion chromatography   A variant of gel filtration in which separation of substances in a mixture is achieved by collecting the fraction of the sample containing the molecules whose size is greater than the exclusion size of the gel beads. (See CHROMATOGRAPHIC TECHNIQUES.)

gel filtration   A technique for separating a substance from others in a mixture by passing the mixture through gel

Gel Filtration

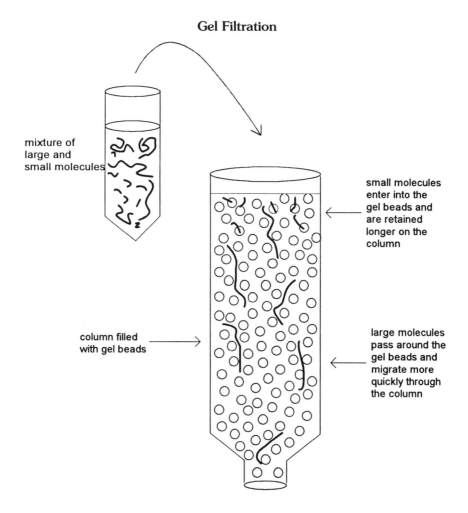

mixture of large and small molecules

small molecules enter into the gel beads and are retained longer on the column

column filled with gel beads

large molecules pass around the gel beads and migrate more quickly through the column

beads in a column. Separation is based on the size of the molecules and depends on the size of the spaces between the polymeric gel molecules (i.e., the pores); substances whose molecules are smaller than the pores enter the gel, so their movement through the gel is slower than larger molecules, which have less tendency to pass through the gel pores but pass around the gel beads. Each type of gel has a characteristic pore size that determines the exclusion size: the maximum molecular size that can enter the gel. Molecules larger than the exclusion size are completely excluded from the gel.

gel retardation assay   A technique for determining the presence of DNA–protein complexes in a given DNA fragment by observing whether the rate of movement of the fragment in an electric field is slowed in the presence of a particular protein.

gel transfer   A term applied to the general process of transferring substances separated by gel electrophoresis from the gel to a membrane for analysis (e.g., for Southern, northern, or western blot analyses). Blotting is one type of gel transfer. (See SOUTHERN BLOT HYBRIDIZATION.)

Genbank   A national database of nucleic acid and protein sequences contributed by various investigators around the world, currently maintained at the National Library of Medicine. The most comprehensive U.S. national database of this type, the Genbank is divided into 13 sequence categories: primate, mammal, rodent, vertebrate, invertebrate, organelle, RNA, bacteria, plant, viral, bacteriophage, synthetic, and unannotated.

gene   A sequence of DNA nucleotides that carries the complete code required for the biosynthesis of a polypeptide.

gene bank   A group of genes that are coordinately controlled.

gene cloning   The science of creating recombinant DNAs that can be inserted into, and copied by, a host microorganism. The term, and the power of the technique, derives from the ability to rapidly grow and easily manipulate large populations of microorganisms carrying the recombinant DNA from a single cell (i.e., a clone).

gene conversion   A mechanism proposed to explain coincidental evolution of duplicated genes (see COINCIDENTAL EVOLUTION) in which a DNA strand from one gene copy becomes paired with the complementary DNA strand from the other gene copy; any area of mismatch (presumably representing a mutation that occurred in one of the gene copies) is then repaired by a MISMATCH REPAIR system.

gene duplication   An error in the normal process by which genes are copied, resulting in a copy of a gene being placed in the same DNA strand as the gene from which it was copied.

gene expression   Any gene activity. Gene expression may include gene transcription into mRNA, translation of mRNA into protein, or activation of a preexisting gene product (protein).

gene families   A group of genes whose nucleotide base sequences show a high degree of sequence homology to one another. In evolution, gene families are believed to arise through gene duplication.

gene flow   The tendency for gene frequency to appear in one population as the result of interbreeding with another population in which the gene is present.

gene frequency   The relative percentage of occurrence of a particular gene relative to all versions (i.e., alleles) of that gene.

gene library   See LIBRARY.

Cloning Genes in Bacteria

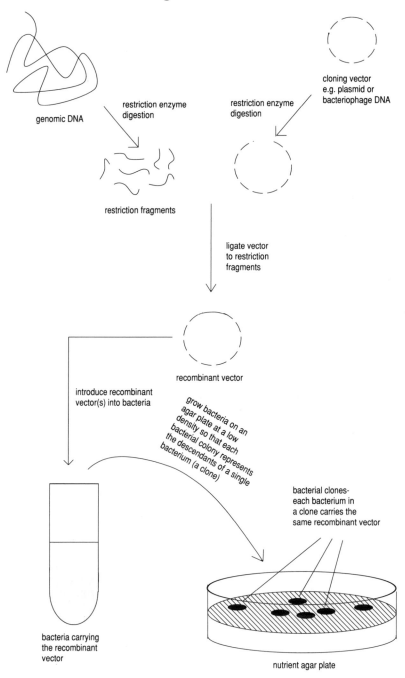

genomic DNA

restriction enzyme digestion

restriction enzyme digestion

cloning vector e.g. plasmid or bacteriophage DNA

restriction fragments

ligate vector to restriction fragments

recombinant vector

introduce recombinant vector(s) into bacteria

grow bacteria on an agar plate at a low density so that each bacterial colony represents the descendants of a single bacterium (a clone)

bacterial clones- each bacterium in a clone carries the same recombinant vector

bacteria carrying the recombinant vector

nutrient agar plate

gene probe See PROBE.

generation time The average time between the appearance of parent and progeny generations in a population.

gene(s), codominant Different versions (alleles) of a particular gene, both of which are active in the heterozygous state. (See HETEROZYGOTE.)

gene(s), dominant A version of a particular gene (allele) whose expression obscures the expression of its recessive allelic form when both are present in the heterozygous state.

gene(s), pseudogenes A variant form of a particular gene that has become permanently inactivated over time as the result of genetic drift.

gene(s), recessive A version of a particular gene (allele) whose expression is obscured by the expression of its dominant allelic form when both are present in an organism (the heterozygous state).

gene splicing A term applied to the general area of recombinant DNA and to the process of SPLICING eukaryotic RNAs. (See EXON, INTRON.)

genetic code The sequence of consecutive nucleotide bases in a strand of DNA or RNA that specifies the sequence of amino acids in a protein or polypeptide. (See CODON, DEGENERATE CODE, and APPENDIX II.)

genetic disease A disease caused by an alteration or mutation in a gene resulting in an aberrant form of the protein coded for by the gene. Because genetic diseases are based on alterations in genes themselves, genetic diseases are transmissible to offspring that receive the faulty gene. Hemophilia and sickle cell anemia are examples of genetic diseases.

genetic drift The process of changes in gene structure in evolution as the result of a random substitutions, loss, or insertions of nucleotide bases in the DNA.

genetic engineering The manipulation of genes through the use of recombinant DNA techniques to modify the function of a gene or genes for a specific purpose.

genetic information Any information carried in terms of the sequence of nucleotide bases in DNA.

genetic map A map of the genome of an organism based on the relative distances between genetic markers. (See LINKAGE MAP.)

genetics The study of the process by which traits are transmitted from parent to offspring; the study of inheritance.

genome All the genetic information carried in the HAPLOID number of chromosomes.

genomic DNA The DNA representing the entire genome. In laboratory terminology, it is used to describe a pure preparation of total native DNA isolated from tissue or a cell culture.

genomic library A library created in a particular vector from genomic DNA such that the entire genome is included in the library.

genotype The set of all genes, including the different alleles, either expressed or not expressed, that is carried in the DNA of an organism.

gentamycin An aminoglycoside antibiotic isolated from *actinomycetes* that is active against a number of gram-positive cocci-type bacteria.

genus A subclassification of organisms between family and species. The standard classification categories in order of general to increasingly detailed biological characteristics is kingdom, phylum, class, order, family, genus, species.

germ cell   A reproductive cell or any cell giving rise to a reproductive cell such as an oocyte or spermatocyte.

germicide   Any agent, chemical or physical, that destroys disease-causing microbes.

germinal centers   A region of the lymph node containing a mass of rapidly dividing B cells. (See B-LYMPHOCYTES.)

germination   The growth of a plant from a spore or a seed.

germline   Embryologically, the cells that will give rise to germ cells in the adult organism.

germline therapy   Gene therapy based on the introduction of new genetic material or alteration of existing genetic material in cells giving rise to either sperm or egg. In germline therapy, the new or altered genetic material can be passed from parent to offspring.

ghost cells   A cell, usually a bacterial or red blood cell, that lacks much or most of its internal contents as the result of lysis (see LYSE) and resealing of the cell membrane. Because they can fuse with other cells, ghost cells have been used as a means of packaging and delivering drugs to other cell targets.

gibberellin   A group of plant hormones that induces growth and maturation in flowering plants.

Gibbs–Donnan effect   The observation that charged molecules on one side of a semipermeable membrane often fail to become evenly distributed on both sides of the membrane. This effect may occur because other charged substances that cannot diffuse across the membrane produce an electric field that influences the migration of charged molecules.

Gibbs free energy   The energy either released or used by a chemical reaction. The total Gibbs free energy (ΔG) is given by $\Delta G = \Delta H - t\ \Delta S$, where ΔH is the energy released (or used) in chemical bond breakage (or formation) during the chemical reaction, ΔS is a measure of the change in entropy (disorder) of the molecules involved in the reaction, t is the temperature at which the reaction occurs.

Gilbert, Walter (b. 1932)   Walter Gilbert became famous as a coinventor of the Maxam–Gilbert technique of DNA sequencing and for his research on the intron–exon structure of eukaryotic genes. He shared the Nobel Prize in chemistry in 1980 with Paul Berg.

glial cell   A cell type occupying the spaces between brain neurons. The five major subclasses are Schwann cells, oligodendrocytes, microglia, astrocytes, and ependymal cells.

glial fibrillary acidic protein (GFAP)   A protein that assembles into a cytoplasmic network of intermediate filaments found only in glial cells.

globin   A group of proteins that forms the subunits of the oxygen-carrying molecules hemoglobin and myoglobin. A mutation in the globin genes is responsible for the oxygen-transporting defect seen in sickle cell anemia.

globotriaosylceramide   A glycolipid molecule that accumulates in patients with Fabry disease, which is caused by a deficiency of the enzyme α-galactosidase A.

globular actin (G actin)   The basic monomeric subunit that polymerizes to form the characteristic actin filaments in muscle. G actin is a single polypeptide of 375 amino acids.

glucagon   A small (29 amino acids) polypeptide hormone that stimulates the breakdown of glycogen into glucose primarily in the liver.

glucoamylase   An enzyme, found largely in saliva but also in the juices of the lower digestive tract, that catalyzes the breakdown of complex sugars (e.g., starch) by cleaving the bonds between two adjacent glucose subunits.

glucocorticoid   One of the three classes of steroid hormones produced by the outer layer (the cortex) of the adrenal gland. The glucocorticoids—cortisone, corticosterone, and cortisol—regulate glucose metabolism and act as anti-inflammatory agents.

glucogenesis   (gluconeogenesis)   The process of creating glucose from its own metabolites. This pathway is active under conditions where the rate of metabolism of glucose is reduced.

glucose   A six-carbon sugar that is the major source of energy for most of the animal kingdom. Energy is generated through the oxidation of glucose to yield carbon dioxide and water (See TRICARBOXYLIC ACID CYCLE.)

glucose effect   The blockage by the presence of glucose of the induction of genes involved in sugar metabolism such as the LAC OPERON in bacteria.

glucose isomerase   The enzyme that catalyzes the conversion of glucose into the structural isomer fructose.

glucose oxidase   An enzyme derived from certain molds that catalyzes the conversion of glucose into gluconic acid with the formation of hydrogen peroxide. Glucose oxidase is widely used for the determination of glucose in the urine in the diagnosis of diabetes since the hydrogen peroxide produced in the reaction can be used to oxidize certain aromatic compounds that form colored products.

glutamic acid   An amino acid whose side chain is

$$-CH_2-CH_2-C\!\!=\!\!O$$
$$\diagdown$$
$$OH$$

The COOH group gives glutamic acid its acidic nature.

glutamine   An amino acid whose side chain is

$$-CH_2-CH_2-C\!\!=\!\!O$$
$$\diagdown$$
$$NH_2.$$

Glutamine plays an important role as an intermediate in the transfer of amino groups in the biosynthesis and degradation of a number of other amino acids.

glutathione   A molecule, made up of three amino acids linked end to end (a tripeptide: glutamate-cysteine-glycine) that acts as a reducing agent. Glutathione plays a role in determining the proper folding of newly synthesized proteins through the cross-linking of cysteine residues.

glycerides (mono-, di-, tri-)   A class of lipids in which one or more fatty acids is covalently attached to a glycerol molecule. Glycerides are divided into mono- (one fatty acid molecule), di- (two fatty acid molecules), and triglycerides (three fatty acids). Glycerides are important as storage vehicles of fat.

glycerol   The simplest carbohydrate containing three carbon atoms with the structure

$$H_2C-CH-CH_2$$
$$|\quad\ \ |\quad\ \ |$$
$$HO\ \ OH\ \ OH$$

Because of its ability to absorb water, glycerol is used commercially as a moisturizer.

glycine   The simplest amino acid, whose side chain consists only of a hydrogen atom.

glycocalyx The cell coat; an outer coating, rich in carbohydrates on the surface of most eukaryotic cells. The glycocalyx also contains some glycolipids and proteoglycans that may form part of the extracellular matrix (ECM).

glycogen A complex storage polysaccharide consisting of branching chains of glucose molecules. Glycogen is the primary source of glucose, which is produced mostly from glycogen breakdown in the liver under conditions where the amount of free glucose is insufficient for the body's needs.

glycolipid A sugar or polysaccharide covalently attached to a lipid. Glycolipids are important components of animal cell membranes. Cerebrosides and gangliosides, which are derivatives of the lipid sphingosine, are important components of glycolipids in membrane receptors in the brain.

glycopeptide A polypeptide covalently linked to a sugar or polysaccharide. Glycopeptides are divided into two classes, depending on whether the sugar(s) are linked to the polypeptide by an oxygen atom (O-linked) or a nitrogen atom (N-linked). (See PEPTIDOGLYCAN.)

glycoproteins Proteins linked to sugars and/or polysaccharides that are prevalent on the outside surfaces of cell membranes. Glycoproteins are components of specialized receptor molecules, the extracellular matrix (ECM) and in the Golgi apparatus of eukaryotic cells.

glycosaminoglycans (GAG) (1) Long branched chains of sugar molecules built from repeating disaccharide subunits containing amino groups. Glycosaminoglycans are present on the surfaces of eukaryotic cells where they are believed to play a role in cell–cell and cell–substrate recognition. (2) Group-specific antigens. The proteins encoded for by the GAG gene of a retrovirus. The GAG proteins are the components of the virus capsid.

glycoside A compound formed between a sugar and some other type of molecule (e.g., a protein, lipid, or other organic molecule).

glycosidic linkage (bond) A covalent bond between the carbonyl-containing carbon of a sugar and another molecule, usually another sugar, protein, or lipid.

glycosylation The process of adding polysaccharides to polypeptides destined to become glycoproteins. Glycosylation takes place primarily in the interior of the ENDOPLASMIC RETICULUM (ER) during the synthesis of the polypeptide to be glycosylated.

glyoxylate cycle A pathway used by plants and bacteria for obtaining energy from acetate and other two-carbon compounds that are metabolized into acetate or acetyl groups. The glyoxylate cycle is similar to the Krebs cycle, with many of the same intermediate steps.

glyoxysome An organelle in plants in which the glyoxylate cycle is carried out.

Golgi apparatus (Golgi body) In eukaryotic cells, a series of membrane-bound vesicles arranged in a stack in which the polysaccharides of glycosylated polypeptides are progressively altered or processed prior to their being sorted and transported to the cell surface.

gonadotrophic hormones (gonadotrophins) A group of polypeptide hormones made by the pituitary gland that stimulate accessory cells surrounding the oocyte to release progesterone, which causes the oocyte to mature. In animals gonadotropins begin to appear at the age of sexual maturity.

gonadatropin The collective name for follicle stimulating hormone (FSH) and lutinizing hormone (LH), small polypeptide hormones made in the anterior

Golgi Apparatus

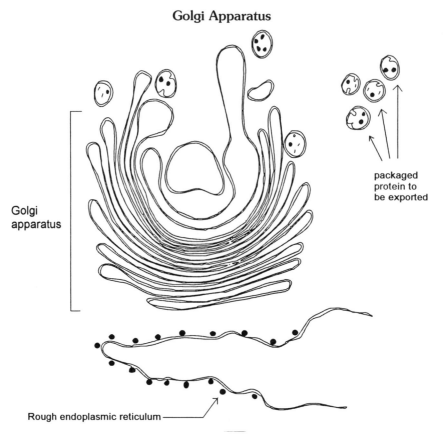

Golgi
apparatus

packaged
protein to
be exported

Rough endoplasmic reticulum

portion of the pituitary gland (adenohy-pophysis) and that act to stimulate the reproductive organs.

Gl phase   A segment of the cell cycle representing the time period between mitosis and the onset of DNA synthesis (S phase).

G protein(s)   A class of cell-mem-brane-bound proteins that bind GTP and/or GDP and act to alter certain met-abolic pathways or gene expression when a specific ligand binds to a receptor on the outside of the cell. (See GUANO-SINE TRIPHOSPHATE, GUANOSINE TRI-POSPHATE PROTEIN.)

graft-vs.-host reaction   A deleteri-ous immune reaction in which lympho-cytes present in grafted tissue attack the tissues of the host.

gram   A universally adopted measure of mass in the scientific world. A gram is defined as one thousandth of the mass of one liter of pure water at a tempera-ture where its density is greatest, namely just above the freezing point (0°C).

gramicidins   A class of polypeptide antibiotics isolated from *Bacillus brevis*. Gramicidins act as ionophores in which ions are carried across the bacterial cell wall in the interior of the circular mole-cule.

Gram stain   A method for staining bacteria developed by Christian Gram in 1884. The bacterial cells are stained with crystal violet, which is preferentially re-tained by the thick cell wall on the sur-faces of gram-positive bacteria, as opposed to gram-negative bacteria, which do not retain the stain. The gram-positive–gram-negative classification system is particularly useful because in a wide variety of bacteria the gram stain shows a correlation with sensitivity to antibiotics.

grana   Stacks of thylakoid disks inside the CHLOROPLAST.

granulocytes   A class of leukocytes composed of neutrophils, eosinophils, and basophils. Granulocytes are active in allergic immune reactions, such as allergic skin lesions and arthritic inflammation.

gratuitous inducer   A molecule that, because it structurally resembles a certain inducer of transcription, can act to induce transcription in lieu of the authentic inducer, such as IPTG in lieu of lactose as a gratuitous INDUCER of the LAC OPERON.

gray matter   That portion of the neural tissue of the spinal cord composing the nerve cell bodies in contrast to the white matter, which is made up of the nerve cell axons and dendrites. In cross section the gray matter is seen as a butterfly-shaped structure running through the interior of the spinal cord.

Griffith, Frederick (1881–1941) A bacteriologist who demonstrated that heat-killed, pathogenic pneumococcus bacteria could transform live, nonpathogenic pneumococci into the pathogenic form when the two were mixed together. This experiment gave rise to the work of Oswald T. Avery, Colin M. MacCleod, and Maclyn McCarty, which showed the transforming factor to be DNA.

griseofulvin   An antifungal agent produced by *Penicillium griseofulvum*. Giseofulvin appears to act by inhibiting the movement of chromosomes during mitosis by interfering with the spindle apparatus. (See MITOSIS.)

group translocation   A type of active transport in bacteria in which compounds entering the cell by passive diffusion are immediately modified (e.g., by phosphorylation) such that they cannot passively diffuse back across the cell membrane. In this way compounds entering the cell are trapped in the CYTOSOL where they accumulate.

growth curve   A graph in which the number of individuals in some population of organisms (e.g., cells in culture, animals in a herd, fish in a pond) is

Three-phase Growth Curve

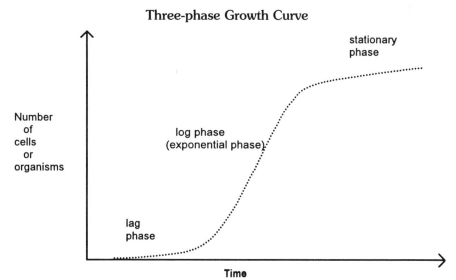

plotted as a function of time. (See GROWTH PHASES.)

growth factors   A group of small secreted polypeptides that bind to receptors on certain specific target cells and stimulate cell division in those target cells. Growth factors have become the focus of intensive research because of their ability to influence the physiology of growth and because many of them bear a close relationship to oncogenes.

growth hormone   A growth factor, produced by the anterior lobe of the pituitary gland, that stimulates the growth of bone and muscle during childhood. Growth hormone was one of the first bioactive factors whose genes were cloned and expressed in transgenic animals, thereby demonstrating the feasibility of curing genetic disease by gene therapy.

growth media   A synthetic solution of nutrients to support the growth of cells or microorganisms in culture. (See DEFINED MEDIUM, MINIMAL MEDIUM.)

growth phases   The different stages of growth of a culture of microorganisms (usually applied to cultures of bacteria) as reflected in the shape of the growth curve. There are three growth phases: (1) a period of slow growth just after the organisms are inoculated into fresh growth medium (lag phase), (2) a period during which the population doubles at a fixed interval of time (exponential phase or log phase), and (3) an indefinite period during which growth is slow or completely stopped as the culture becomes overcrowded (stationary phase).

growth rate   The change in the number of organisms in a population divided by the length of the time interval over which the change in the population number took place. For example, a culture of cells in which there were 2500 organisms on one day and 7000 organisms three days later has a growth rate of $(7000 - 2500)/3$ or 1500 organisms/day.

Grunstein and Hogness method   The technique of HYBRIDIZATION of a DNA probe to whole, lysed, bacterial colonies that have been transferred, by blotting, onto a nitrocellulose filter or other hybridization membrane; colony hybridization.

GT-AG rule   The observation that INTRON sequences in DNA always begin with GT as the first two nucleotides and always end with AG as the last two nucleotides. The GT-AG rule plays a role in the mechanism of RNA splicing by which messenger RNA (mRNA) is created from heterogeneous nuclear RNA (hnRNA).

guanine   One of the four PURINE bases normally found in DNA and RNA.

guanosine   A ribonucleoside consisting of guanine and the ribose sugar.

guanosine, 7-methyl   A derivative of the normal nucleoside guanosine, found in the cap of eukaryotic mRNAs. (See CAPPED 5'-ENDS.)

granosine diphosphate (GDP)   A derivative of guanosine monophosphate formed by the addition of another phosphate group so that two phosphate groups are attached to the fifth carbon atom of the ribose sugar.

guanosine monophosphate (GMP)   The ribonucleotide containing guanine; a derivative of guanosine formed by phosphorylation of the fifth carbon of the ribose.

guanosine triphosphate (GTP)   Guanosine with three phosphate groups attached to the fifth carbon atom of the ribose sugar. GTP, together with ATP, CTP, and UTP, is a direct precursor of RNA and a high-energy compound that provides energy to drive other biochemical reactions.

guanosine triphosphate binding protein (G protein)   A membrane-

Guanosine, Deoxyguanosine, Dideoxyguanosine

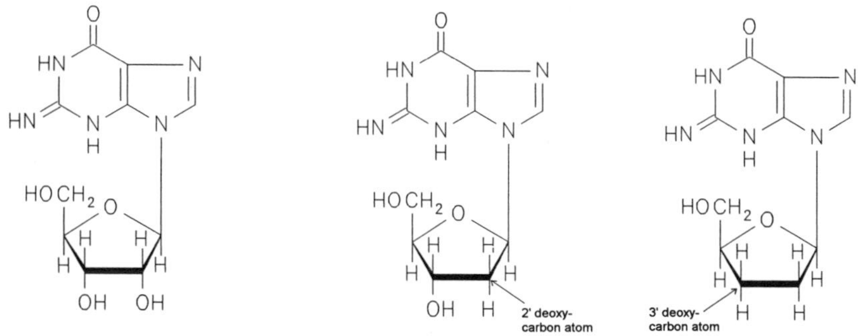

guanosine

deoxyguanosine

dideoxyguanosine

bound protein that acts as an intermediate between the binding of a stimulatory molecule such as a hormone to a receptor on the outside of the cell and the biochemical effect that ultimately takes place inside the cell. The mechanism by which G proteins transmit the signal (i.e., the binding of the stimulatory molecule to its receptor) from the outside of the cell to the inside is not entirely understood but is known to involve the exchange of a molecule of GDP for a molecule of GTP by the G protein as part of the process.

gut associated lymphatic tissue (GALT)   Patches of lymphoid tissue in the small intestine; Peyer's patches.

H

habituation   The tendency of some neurons to require longer than normal refractory phases or stimulation by stronger than normal nerve impulses in order to trigger an action potential if action potentials have been triggered in that neuron in the recent past.

hairpin loop   The folding back of a nucleic acid strand on itself. Hairpins are created by internal base pairing between purine and pyrimidine bases along two separate segments of the nucleic acid. (See LOOPED DOMAINS.)

half register   In repeated sequences in which the repeat unit can be divided into two halves, half register refers to a misalignment of the two chromosomal copies such that the first half of a repeat unit is aligned with the second half of the other chromosomal copy. For example, if the two halves of the repeat units are designated X and Y, then the repetitive portion could be represented by
. . . XYXYXYXYXYXYXYX . . .
which is in half register with
. . . . XYXYXYXYXYXYXYX. . . .

halophile   Salt-loving; a type of bacterium that can grow in high concentrations of sodium chloride (NaCl).

haploid   The state of a cell having only one set of alleles as opposed to the diploid state in which a cell normally contains two copies of each ALLELE.

haploid number   Having one-half the number of chromosomes normally present in a diploid cell, thus being in the haploid state (e.g., in gametes).

haplotype A particular set of markers (e.g., RFLPs or alleles in a certain region of a chromosome). The term was originally applied only to clusters of alleles in the major histocompatibility complex (MHC) but is now applied to any specified genetic locus.

hapten A small molecule that can act as an immunogen only when combined with a larger molecule.

Hardy–Weinberg law A mathematical formulation describing how two different alleles are distributed among the individuals in a population. For two allelic forms, D and d, of a certain gene, assuming that (1) mating between individuals is random, and (2) if the frequency of a certain allele D in the population is p and the frequency of the d allele form is q, then (i) the fraction of individuals homozygous for D (DD) is always p^2; (ii) the fraction of individuals HETEROZYGOUS for D (Dd) is always $2pq$; and (iii) the fraction of individuals HOMOZYGOUS for d (dd) is always q^2.

harvesting Collection of cells, organisms, or growth medium from an experimental population, generally for purposes of analysis or extraction of biochemicals.

HB101 A substrain of the bacterium, *Eschericia coli* widely used as a host in which to grow recombinant vectors because of its high efficiency of DNA uptake and transformation.

heat shock genes A set of genes found throughout the animal kingdom that are suddenly and rapidly transcribed in a coordinated fashion when cells are subjected to certain kinds of stress, such as a sudden rise in temperature.

heat shock response element (HSE) A certain nucleotide sequence in the PROMOTER of the heat shock genes

(CNNGAANNTCCNNG). The binding of a transcriptional enhancer protein to this sequence is the first event in the activation of the heat shock genes.

heavy chain (immunoglobulin heavy chain) The longer of the two peptide chains that make up an antibody molecule (e.g., IgG).

HeLa cell A line of tissue culture cells derived from a cervical cancer by Gey, Coffman, and Kubicek in 1951. The cell line designation is derived from the name of the tumor donor and was the first epithelial-like cell derived from human tissue to be placed into tissue culture.

helicase An enzyme(s) that catalyzes the unwinding of the DNA helix during DNA replication at a point just ahead of the replication fork.

helix, α Refers to one type of three-dimensional conformation that a protein assumes in the cell. An α helix is formed due to the formation of many hydrogen bonds between nearby amino acids in the protein. Hydrogen bonds form between every three amino acid residues, thus providing regularity to the helix structure. Another conformation of proteins is the β-PLEATED SHEET.

helper T cell A specialized type of T cell whose function is to stimulate other T-lymphocytes (e.g., cytotoxic T cells), which then carry out various immune functions. Helper T cells are stimulated to divide after they are exposed to a foreign MHC antigen, which is presented to it by specialized antigen-processing cells; the stimulatory activity of T helper cells is mediated by INTERLEUKINS. (See MAJOR HISTOCOMPATIBILITY COMPLEX.)

helper virus A virus that provides a critical function to a defective virus when they both simultaneously infect the same cell. The oncogene-carrying retroviruses are examples of replication-defective viruses that can grow only when coinfected with the normal WILD-TYPE counterpart, which does not carry an oncogene.

hemagglutination inhibition assay An assay for the presence of a hemagglutinating virus or other antigen by observing the loss of the ability of a test sample to agglutinate red blood cells after being treated with an antibody against the agglutinin whose presence is suspected.

hemagglutinin Any substance that can cause red blood cells (RBCs) to agglutinate by binding to certain sites on the RBC membrane. Because the clumping (agglutination) of RBCs is easily seen even in the presence of small amounts of hemagglutinin, hemagglutination has been widely used as an assay for the presence of certain hemagglutinating viruses or other antigens.

hemagglutinin - neuraminidase protein (HN protein) A protein, derived from the coat of the paramyxovirus (see SENDAI VIRUS), that binds strongly to the cell membranes of many types of animal cells. For this reason, the HN protein is used to create liposomes intended to deliver agents (e.g., therapeutic agents) to various animal cells (fusogenic vesicles).

heme An organic, iron-containing, ring-shaped molecule that is the oxygen binding group in hemoglobin.

hemicellulose The name given to a mixture of long polysaccharides with a cellulose-like structure that, together with pectin, forms an amorphous matrix in which the cellulose fibrils of the plant cell wall are embedded.

hemidesmosome Literally meaning "half desmosome," a specialized membrane junctional complex in the epithelial cell membrane. It structurally resembles a DESMOSOME but, unlike a desmosome, is present at the site where

an epithelial cell makes contact with the basal lamina.

hemizygous The cellular state of having one copy of a gene in a genome that is normally diploid for all genes. The term always refers to a particular gene or group of genes; for example, a cell is said to be hemizygous for gene x.

hemoglobin The large bloodborne molecule carrying oxygen and carbon dioxide between the lung and tissue. Hemoglobin consists of a heme group and four polypeptide chains, namely two α-globin chains and two β-globin chains.

hemolymph The fluid in the body cavity of insects serving the same gas-exchange functions as blood.

hemolysins A group of bacterial toxins that cause hemolysis by attacking red blood cell membranes.

hemolysis The lysis (breaking open) of red blood cells.

hemolytic plaque assay An assay, based on the localized hemolysis of red blood cells (RBCs), that appears as a plaque when the RBCs are spread out in a layer of agar. The hemolytic plaque assay is used to demonstrate the secretion of specific antibodies by antibody-producing cells mixed with the RBCs.

hemophilia A genetic disease based on the inability of the afflicted individual to make a critical component (factor VIII) of the blood-clotting system. As a result, even minor cuts or bruises may result in dangerous, uncontrolled internal hemorrhage or bleeding.

hemopoiesis The generation of red blood cells by cell division of certain stem cells in the bone marrow.

Henderson–Hasselbalch equation A mathematical formulation that governs the pH of a given buffer solution: If pK is the negative logarithm of the equilibrium constant (K_d) for the ionization of the acid form of the compound used to buffer the solution for the reaction HA $\leftrightarrow$ H$^+$+A$^-$ and [HA] and [A$^-$] are the molar concentrations of the un-ionized and ionized forms of the buffer, respectively, then pH=pK+log ([A$^-$]/ [HA]).

heparin   A sulfated glycosaminogly-can used medically to block blood clot-ting. The anticoagulant activity of heparin is based partly on the strong binding of the heparin molecule to anti-prothrombin III, a blood protein that plays a critical role in the blood-clotting pathway.

hepatitis   An inflammatory disease of the liver characterized by severe, chronic jaundice, caused by accumulation of liver by-products, and by general malaise.

hepatitis virus   The viral agent, a small DNA virus, that causes the form of hepatitis that infects many people in certain areas of the world. The three subclasses are A,B, and C.

heptad repeat   A tandemly repeated segment of seven amino acids in certain proteins. The heptad repeat is found in virtually all intermediate filament pro-teins throughout the animal kingdom.

herbicide   A chemical agent toxic to plants.

hereditary disease   See GENETIC DIS-EASE.

hereditary persistence of fetal he-moglobin (HPFH)   A genetic state in which the normal adult δ- and β-hemoglobin genes are absent, and so the fetal hemoglobin genes continue to be expressed past the time when they would normally be turned off.

heredity   The study of how physical traits are transmitted from parent to off-spring: the study of inheritance. (See MENDEL'S LAW.)

herpes   A family of large DNA viruses that infects humans and produces both acute infections, such as chickenpox, and those resulting from persistent, la-tent infections, such as shingles, caused by the same virus, *herpes zoster*. The members of the herpes family of viruses are herpes simplex, *Herpesvirus simiae, varicella-zoster,* cytomegalovirus, and Epstein–Barr virus.

herpes simplex virus (HSV)   A member of the herpes family of viruses implicated as the causative agent in some cervical cancers.

Hershey–Chase experiment   A classic experiment conducted by Martha Hershey and Alfred Chase in 1952 that demonstrated that DNA was the heredi-tary material. In their experiment, bacte-riophage containing ^{32}P-DNA and ^{35}S-protein were allowed to attach to host bacteria. When, after several minutes, the attached bacteriophage were stripped from the bacteria by strong me-chanical agitation, it was found that the ^{32}P label, not the ^{35}S label, had entered the host cells.

heterochromatin   A very condensed form of CHROMATIN seen in the nucleus during interphase and found to be a tran-scriptionally inactive form.

heteroduplex   Base pairing between nucleic acid strands from different sources, for example, RNA and DNA, or DNA from two different species.

heteroduplex mapping   A tech-nique for determining the location of a particular sequence of nucleotide bases along a segment of a nucleic acid by creating a heteroduplex between the nu-cleic acid to be mapped and a reference nucleic acid strand.

heterofermentation (heterolactic fermentation)   A type of fermenta-tion characteristic of enteric bacteria *(En-terobacteriaceae)* in which only part of the fermentation product is lactic acid and the other part is formate and acetyl CoA.

heterogeneous nuclear RNA (hnRNA) The general name give to all unprocessed RNAs in the nucleus. The name derives from the great heterogeneity in size and type of RNA present in the nucleus before the RNAs are processed and transported to the cytoplasm.

heterokaryon A multinucleated hybrid cell created either from the fusion between two or more cells or by cell division without CYTOKINESIS.

heterolactic fermentation Heterofermentation.

heterotroph See CHEMOORGANOTROPH.

heterozygote An individual who is heterozygous for a particular gene.

heterozygous The state of containing two different ALLELES of a gene. For example, that a cell or an individual is heterozygous for the trait that causes sickle cell anemia if the normal and abnormal copies of the β-globin gene are present.

hexose Any six-carbon sugar.

H-2 histocompatibility The match of tissue proteins that, in the mouse, determines whether a tissue graft will be rejected by the immune system of the host. H-2 compatibility is determined by a large gene complex that codes for cell surface glycoproteins. (See MAJOR HISTOCOMPATIBILITY COMPLEX.)

high frequency recombination strain (Hfr) Certain strains of bacteria which can transfer and donate their genes to recipients at a high rate. The high frequency of transfer of their chromosomes and subsequent recombination is based on the presence of an integrated F-factor that allows mating to occur between neighboring bacteria. See CONJUGATION.

highly repetitive DNA A fraction of the genomic DNA in eukaryotic cells consisting of short sequences repeated thousands of times throughout the genome and found to be equivalent to satellite DNA

high-mobility group protein (HMG protein) A heterogeneous group of proteins of unknown function, they are part of the CHROMATIN but are not HISTONES.

high-performance liquid chromatography (HPLC) A technique utilizing a variety of separation techniques, including those used in ion-exchange, size-exclusion, or reverse-phase chromatography but that achieves a higher level of resolution of separation by incorporating improvements in the packing of columns and flow of solvents through columns under high pressure. By this method extremely sharp elution peaks of substances from the column can be obtained. (See CHROMATOGRAPHIC TECHNIQUES.)

Hill reaction The light-energy-harvesting reactions in photosynthesis in which light energy is stored in the form of high-energy electrons. Discovered by Robert Hill in 1939.

Hind III A restriction enzyme whose recognition sequence is
5′ AAGCTT 3′
3′ TTCGAA 5′

histamine A substance stored in the granules of mast cells and released during allergic response. Histamine release causes smooth muscle contraction, secretory activity in mucous epithelium, and other symptoms of allergic reaction.

histidine An amino acid whose side chain is

$$-CH_2-C=CH$$
$$\quad\quad\quad | \quad\quad |$$
$$\quad\quad\quad N \quad NH$$
$$\quad\quad\quad \backslash\backslash \quad |$$
$$\quad\quad\quad\quad\quad CH$$

histocompatibility   The state of similarity, or dissimilarity, between the proteins of a grafted tissue and proteins of the host on which the tissue is grafted. The degree of histocompatibility is the major factor in determining whether a host will accept or reject a tissue graft.

histones   A group of proteins tightly associated with DNA to form structures known as nucleosomes. The five subgroups are H2A, H2B, H3, H4, and F1. They appear to play a role in regulating the expression of genes.

Hodgkin's disease   A type of cancer of the tissue of the lymphatic system including the lymph nodes, spleen, tonsils, and thymus gland. The disease is characterized by fever, lymph node enlargement, and weight loss.

Holliday junction   The linkage of two homologous double-stranded DNAs by ligation of the broken end of one strand with the broken end of a strand on the homologous DNA.

holoenzyme   The complete and functional form of an enzyme; the polypeptide portion of enzyme plus any other factors necessary for normal enzymatic activity (e.g., coenzymes and cofactors).

homeobox   Those DNA sequences found in HOMEOTIC GENES of the fruit fly, *Drosophila melanogaster,* and in amphibians and mammals. Homoebox sequences are expressed in early development and appear to play a role in limb and appendage development similar to the role of the homeotic genes.

homeostasis   The state of being in balance. In biological systems, the term describes a group of biochemical reactions that act together as a system of checks and balances to prevent "overreaction" on the part of any reaction.

homeotic genes   A cluster of genes that determine limb and appendage development in the fruit fly, *Drosophila melanogaster.* Homeotic genes have

Holliday Junction

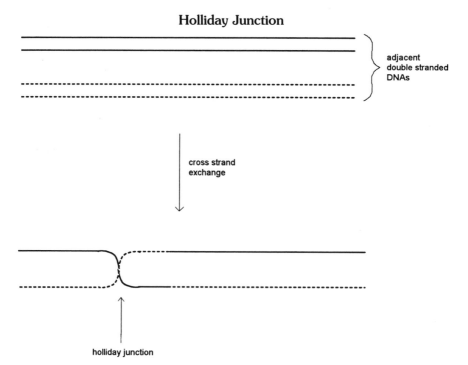

adjacent
double stranded
DNAs

cross strand
exchange

holliday junction

been defined in terms of mutations at certain genetic loci that may cause one type of appendage to develop in place of another (e.g., insect leg in place of an antenna). (See HOMEOBOX, SEGMENTATION, SEGMENTS.)

homeotic selector genes The genes of the bithorax and antennapedia complexes taken together.

homofermentation (homolactic fermentation) Fermentation characteristic of lactic bacteria in which sugar is oxidized by the fermentation pathway to a single product, lactate.

homogenate A crude slurry resulting from the disruption of cell–tissue structure by mechanical means, such as grinding.

homogeneously staining region (HSR) A region of a chromosome containing multiple copies of a particular gene and identified as a segment that stains homogeneously (as opposed to showing a number of bands, as is normally true), with certain stains used to visualize chromosomes under the microscope. The amplification of the dihydrofolate reductase gene (DHFR) in cells exposed to the anticancer agent methotrexate is seen as a homogeneously staining region of giant chromosomes.

homolactic fermentation See HOMOFERMENTATION.

homologous chromosomes Two chromosomes, generally one from each parent, identical in terms of the genes they carry but differing in terms of the ALLELES of those genes.

homologous recombination Recombination between two DNAs at certain sites, where the two DNAs show sequence homology to one another.

homology In general, meaning similar, but not necessarily identical, to

something. In molecular biology it is generally taken to mean "sequence homology."

homopolymer In general, meaning any polymeric molecule containing a single type of monomer. In molecular biology, it refers to a short nucleic acid segment consisting of a single type of nucleotide (e.g., oligo dT).

homopolymer tailing The technique of attaching a nucleic acid homopolymer to the end(s) of a piece of DNA, generally as an intermediate step in cloning. Homopolymer tailing has been widely exploited as a means of cloning cDNAs by attaching homopolymer tails to the ends of the cDNA, which can then be easily annealed to complementary homopolymer tails attached to the ends of a vector molecule. (See ANNEAL, COMPLEMENTARY DNA.)

homoserine The precursor molecule from which, in plants and bacteria, the amino acids methionine, threonine, and isoleucine are made.

homozygous The state in which the genome contains two copies of the same ALLELE of a gene. Organisms are said to be homozygous for a gene.

hormone Any molecule made and secreted by a specific tissue and that causes or induces a specific action or behavior in another (target) tissue (e.g., FSH, LH, or the steroid hormones estrogen and testosterone).

horseradish peroxidase (HRP) An enzyme that causes the breakdown of peroxides by catalyzing the transfer of hydrogen atoms to an oxygen atom on the peroxide. The activity of the enzyme has been exploited as a technique for labeling proteins and nucleic acids colorimetrically instead of by radiolabeling. For this purpose the molecule is attached to an HRP molecule, which is then made visible by using a substrate (S) that becomes colored when it is oxidized by HRP:

$$H_2O_2 + \text{Substrate-}H_2 \xrightarrow{\text{peroxidase}}$$
$$\text{(uncolored)}$$
$$2H_2O + \text{Substrate}_{ox}$$
$$\text{(colored)}$$

host cell   Any cell infected by a virus is referred to as the host cell for that virus.

host range   The group of all cell types susceptible to infection by a virus.

host restriction-modification   Restriction enzyme systems developed by bacteria that inactivate the DNA of infecting bacteriophage by cleaving them with restriction enzymes produced by the bacterial host while protecting the host DNA from cleavage by the same restriction enzyme through another system modifying the host DNA, usually by methylation. (See RESTRICTION ENDONUCLEASE.)

host-vector system   A combination of host cell and virus vector to be used for cloning—for example, a particular strain of the bacterium. *E. coli* (host) and a particular strain of bacteriophage λ (vector) that grows especially well in the host cell.

hotspot   A genetic locus particularly prone to spontaneous alteration or mutation.

housekeeping genes   A vernacular term to describe the genes necessary for basic cell functions required by, and therefore expressed in, all cells.

H-ras gene   The human proto-oncogene homologue of the *ras* oncogene carried by the Harvey sarcoma virus.

human chorionic gonadotropin (HCG)   An ovary-stimulating hormone produced in the placenta after the embryo implants into the wall of the uterus. HCG is referred to as the "pregnancy hormone" because antibodies to HCG are used to diagnose pregnancy.

Human Genome Project   An extensive project, sponsored by the National Institutes of Health, that has as its goal the determination of the complete nucleotide sequence of the entire human genome of about 3×10^9 base pairs of DNA. The project is intended to be carried out at several large research centers throughout the United States. At present, funding of the massive project is still the topic of much debate centered mainly on the ultimate value of such a project given its enormous costs.

human growth hormone (somatotropin)   A polypeptide hormone, produced by the anterior pituitary, which stimulates the liver to produce somatomedin-1. The hormone in turn causes growth of muscle and bone.

human immunodeficiency virus (HIV)   The retrovirus that causes acquired human immunodeficiency syndrome (AIDS).

human leukocyte-associated (HLA) antigens   The proteins of the major histocompatibility locus in the human.

human T-cell leukemia virus   A group of human retroviruses that cause human T-cell leukemias.

humoral antibody   Secreted antibody produced by B cells circulating in the blood. Humoral antibodies mediate the immune response to soluble antigens, as opposed to graft rejection.

hyaluronic acid   A type of glycosaminoglycan in which the repeating disaccharide consists of the sugars glucuronic acid and N-acetylglucosamine.

hybrid   An organism that is the offspring of parents of different GENOTYPES.

hybrid-arrested translation A technique for determining the identity of a cDNA based on hybridizing the cDNA to its corresponding mRNA and then observing the loss of the ability of the mRNA to be translated into protein in vitro. (See COMPLEMENTARY DNA, HYBRIDIZATION.)

hybrid cell A cell produced by cell fusion but that, after several cell divisions following fusion, now contains one nucleus with chromosomal material from both original parent cells.

hybrid dysgenesis A term applied to the inability of certain strains of the fruit fly, *Drosophila melanogaster,* to interbreed because offspring resulting from matings between the strains are sterile or show high frequencies of mutation.

hybridization The formation, in vitro, of a double-stranded nucleic acid segment by hydrogen bonding between two single strands. Experimental use of hybridization is the basis of DNA probe technology, including Southern and northern blot analyses, primer annealing, heteroduplex analysis, and so on.

hybridization probe Any labeled nucleic acid segment used in any of a variety of assays based on hybridization of the labeled nucleic acid to a target nucleic acid.

hybridization stringency A term used to describe the degree of mismatch tolerated by a specific set of hybridization conditions. Hybridization stringency is usually given in terms of the minimal percent base match required for duplex formation between the hybridization probe and the target nucleic. Thus, the chemical and physical conditions under which a hybridization occurs can be adjusted so that the level of homology between the probe and the target is 85%, 90%, and so on. Levels of homology below about 70% are generally considered to be nonhomologous, and so hybridization conditions permitting duplex formation between nonhomologous nucleic acids are called nonstringent.

hybridoma An immortalized antibody-secreting cell created by fusing a myeloma cell to lymphocytes from the spleen of an animal that has been immunized to a particular antigen. Antibody secreting hybridomas are the source of monoclonal antibodies.

hybrid vigor The state in which an offspring is genetically more robust and/ or better equipped for survival than either parent as the result of heterozygosity—that is, receiving a beneficial combination of traits from its parents.

hydrocarbon Any organic molecule comprising only hydrogen and carbon. The most common hydrocarbons are derived from a linear chain of carbon atoms: for example, methane, ethane, propane, and so on, called saturated hydrocarbons, ethylene, propylene, and so on, called unsaturated hydrocarbons, and benzene, toluene, and so on, called aromatic hydrocarbons.

hydrogen bonds Electrostatic attractions between positively charged hydrogen atoms and negatively charged atoms on other parts of a molecule or on other molecules. Hydrogen bonds are the major forces that stabilize the structures of many proteins and the DNA double helix.

hydrolase The general class of enzymes that catalyzes reactions involving hydrolysis.

hydrolysate The product of the hydrolysis of a material (e.g., a protein hydrolysate).

hydrolysis The breakage of any covalent bond by insertion of a water molecule across the bond, for example

$$
\begin{array}{ccc}
-C{=}O & & -C{=}O \quad\ NH- \\
\ |\quad\ \ \xrightarrow{\ H_2O\ } & \ |\quad\ \ +\quad | \\
NH- & & OH \qquad\ H
\end{array}
$$

The carbon–nitrogen bond (the peptide bond in a protein) is said to have undergone hydrolysis.

hydronium ion   A water molecule to which a hydrogen ion is attached: $H^+ + H_2O \rightarrow H_3O^+$. Hydronium ions are the form in which hydrogen ions are normally carried in aqueous solutions.

hydropathy plot   A graph showing the degree of hydrophobicity of each amino acid in a polypeptide as a function of its location in the polypeptide. Hydropathy plots are often used to visualize the clustering of hydrophobic amino acids. If such clustering is observed, it may indicate that the polypeptide in question is a transmembrane protein with the hydrophobic cluster representing a transmembrane domain.

hydrophilic   The property of an atom, molecule, or molecular group having an electrostatic attraction to water molecules. Hydrophilic groups tend to be soluble in water.

hydrophilic signaling molecule   A large class of highly water soluble molecules that, because of their solubility, can diffuse easily across an aqueous medium between a cell from which they are secreted to a target cell where they trigger some specific event. Various growth factors and hormones are commonly encountered as hydrophilic signaling molecules.

hydrophobic   The property of an atom, molecule, or molecular group having no, or very little, electrostatic attraction to water molecules. Hydrophobic groups tend to be insoluble in water.

hydroponics   The science of growing plants in a synthetic, aqueous nutrient medium.

hydrops fetalis   A type of thalassemia in which all four α-chains missing in the hemoglobin molecule are missing as a result of a defect in the DNA that codes for these proteins. Infants carrying the defect almost inevitably die at, or before, birth.

hydroxylapatite   A form of calcium phosphate ($CaPO_4$) used for separation of single- and double-stranded nucleic acids by column chromatography in which the double-stranded form is preferentially retained on the hydroxylapatite matrix. (See CHROMATOGRAPHIC TECHNIQUES.)

5-hydroxymethylcytosine   An unusual derivative of cytosine used in place of cytosine in the DNA of the BACTERIOPHAGE T4. This base protects the T4 DNA from nucleases produced by the bacteriophage during its growth cycle.

hydroxyproline   A derivative of the amino acid proline that is found in collagen and helps to stabilize the collagen molecule. Because the formation of hydroxyproline depends on vitamin C, a deficiency of the vitamin is manifest by a weakening of the collagen fibers, resulting in the skin lesions characteristic of scurvy.

hyperchromicity, hypochromicity   The change in the optical density of a solution of a nucleic acid upon denaturation or renaturation. Denaturation of double-stranded DNA to the single-stranded state results in an increase in the optical density of the sample (hyperchromic shift) at 260 nm. A reduction in optical density (hypochromic shift) accompanies renaturation of single strands to the double-stranded form.

hyperimmune   A state in which an extreme immune response is provoked by an antigen present in quantities normally not effective in stimulating the immune system.

hypermutable phenotype   Bacterial strains lacking the ability to remove uracil molecules that aberrently arise in the cell DNA in place of cytosine. The persistence of uracil results in high rates of mutation in strains carrying this deficiency.

hyperproliferation A state in which cell division occurs at a greater than normal rate.

hypersensitive site, DNase I A region of the CHROMATIN at which the DNA is accessible to the enzyme DNase I. Experimental evidence has shown that many of these sites are places where gene activity is regulated.

hypervariable region A region of the immunoglobulin gene showing a high degree of variability in sequence from one antibody to the next.

hypha A long branching filament of connected cells in fungi. Hyphae may be segmented, with individual nuclei separated from one another by a cell wall, or nonsegmented, with many nuclei sharing a common cytoplasm (multinucleated).

hypochromicity See HYPERCHROMICITY, HYPOCHROMICITY.

hypoxanthine A purine intermediate in the degradative pathway of adenosine. Hypoxanthine can also serve as a precursor for nucleic acid synthesis by a series of reactions known as the salvage pathway.

hypoxanthine-aminopterin-thymine (HAT) medium A type of cell culture growth medium used for negative selection (i.e., HAT selection) of certain kinds of mutant cells that cannot utilize hypoxanthine and/or thymine to make nucleic acids.

hypoxanthine-aminopterin-thymine (HAT) selection The pro-

Hypoxanthine

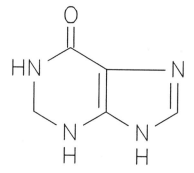

cedure for selecting cells in HAT medium, based on the principle that only those cells that can utilize hypoxanthine and thymine, supplied from outside the cell to make their nucleic acids, will survive in the presence of the drug aminopterin or other folate antagonists that prevent cells from synthesizing their own purine and pyrimidine nucleotides.

hypoxanthine-guanine phosphoribosyl transferase (HGPRT) An enzyme that catalyzes a major step in the formation of ATP and GTP from guanine. This pathway is the only means by which guanine or other purine analogs can enter nucleic acids. Thus, as is true for thymidine kinase in the pyrimidine pathway, manipulation of this enzymatic step provides an important experimental tool for studying gene action by the incorporation of modified bases into DNA.

hypoxanthine-guanine phosphoribosyl transferase (HGPRT) marker Refers to the gene that codes for the HGPRT enzyme as a selectable marker. Cells containing mutant HGPRT genes are resistant to purine derivatives, which are toxic compounds because they become incorporated into DNA via the HGPRT-dependent pathway. (See HYPOXANTHINE-GUANINE PHOSPHORIBOSYL TRANSFERASE.)

I

Ia antigen   Proteins encoded by the I locus of the mouse H-2 histocompatibility complex. (See MAJOR HISTOCOMPATIBILITY COMPLEX.)

idiotope   An antigenic peptide sequence located on the IgG antibody molecule near the antigen binding site; a specific idiotope is associated with a specific antigen binding site so that the same number of idiotopes exist as there are different antibodies.

idiotype   The set of idiotopes on an IgG molecule.

i gene   The bacterial gene that codes for the LAC OPERON repressor protein.

illegitimate recombination   A rare event in which recombination occurs between two DNAs at an apparently randomly chosen site(s) which exhibit no technology to each other.

imaginal disc   Disc-shaped structures symmetrically located on either side of the embryos of fly larvae. These discs give rise to certain adult structures (e.g., legs, eyes, wings).

immortalized cells   Cells that continue to divide indefinitely in tissue culture. Immortalization is a defining property of transformed cells.

immune response   The proliferation of specific antibodies or cells of the immune system, such as macrophages and T- and B-lymphocytes, in response to a foreign antigen.

immune system   The collection of all the cells and tissues (thymus, spleen, lymphocytes, etc.) involved in providing an immune response.

immunoadsorbent   A solid matrix to which antibodies are attached and which is used to purify the antigens from a biological preparation by allowing the antigens to bind to the matrix-bound antibodies.

immunoaffinity chromatography   A technique for purifying antigens by passing a biological preparation or extract over a column containing an immunoadsorbent.

immunoassay   Any technique for determining the quantity of an antigen based on the binding of the antigen to its specific antibody. (See COMPLEMENT-FIXATION TEST, HEMAGGLUTINATION-INHIBITION ASSAY, RADIOIMMUNOASSAY.)

immunoblotting   A technique for determining the presence and properties of an antigen by reaction of labeled antibodies to the antigen after the antigen has been separated according to size and/or charge by gel electrophoresis and then transferred to a membrane.

immunodeficiency   The state of impairment of the immune system resulting in inability or lowered ability to mount an immune response to a cell or particular antigen.

immunodiffusion   A technique for determining the presence of an antigen by allowing an antigen and an appropriate antibody to diffuse into a gel where an immune precipitate forms at the point where antigen and antibody meet.

immunoelectrophoresis   A variation of the immunodiffusion technique in which the antigen is subjected to electrophoresis in a gel, which is then used for assay by immunodiffusion.

immunofluorescence   A technique for visualizing structures in a cell or tissue through the use of antibodies attached to a fluorescing label that bind to antigens in the target structures.

immunogen Any substance capable of provoking an immune response.

immunogenicity The property of being an immunogen—that is, the property of being capable of provoking an immune response.

immunoglobulin (Ig) Any of the globular serum proteins secreted by cells of the immune system for the purpose of dealing with foreign antigens. Immunoglobulins are divided into five classes: IgM, IgG, IgA, IgD and IgE.

immunoglobulin gene switching Developmental changes in the class of immunoglobulin (such as from IgM to IgG) produced by a single lymphocyte as the result of the expression of different genes. Gene switching is accomplished by recombination and changes in the way RNA is spliced so that the products of different portions of the immunoglobulin genes are fused to one another in different arrangements.

immunolabeling The technique of labeling molecules and/or biological structures through the use of antibodies bound to other molecules that serve as labels for the antibody-antigen complex.

immunology The study of the immune system.

impermeable junction A term to describe any cell-cell junctional complex that connects cells so that even small

IgG Type Immunoglobulin

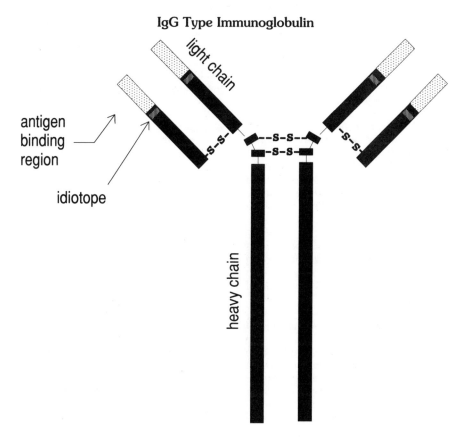

molecules cannot diffuse between the connected cells. Impermeable junctions are generally synonymous with tight junctions.

inclusion bodies   Clumps of material that accumulate in the nucleus or cytoplasm of virus-infected cells. Inclusion bodies consist of aggregates of viral structural components such as VIRION proteins.

indirect-end labeling   A technique for demonstrating the unique nature of a DNA fragment by hybridizing the fragment to a probe representing an end piece of the sequence it is proposed to represent. For example, if it is supposed that the nucleosome-binding DNA sequence is a unique sequence, then a probe may be constructed from the end portion of this sequence, and any DNA isolated from a NUCLEOSOME should hybridize to the probe. (See HYBRIDIZATION, NUCLEOSOME PHASING.)

inducer   Any agent, chemical or physical, that brings about the expression of a gene or gene cluster.

inducible enzyme   An enzyme whose expression requires the presence of a specific inducer.

induction, gene   Transcription of a gene(s) brought about in the presence of a specific agent referred to as the inducer.

induction, phage   The production of lytic BACTERIOPHAGE in bacteria carrying a LYSOGENIC PROPHAGE brought about by treatment with some chemical of physical agent.

informative meiosis   A mating that generates a crossing over between two genetic markers so that linkage between the two markers can be determined. Informative meioses have been used in the genetic analyses of genetic disease such as cystic fibrosis to establish linkage between RESTRICTION FRAGMENT LENGTH POLYMORPHISMS (RFLPs) known to be associated with the disease.

infrared (IR) spectroscopy   An analytical technique for determining the chemical structure of an unknown by observing how much infrared radiation is absorbed by a sample of the unknown at different wavelengths in the infrared spectrum (>1000 nm). The absorption of radiation in this region of the spectrum reflects, and is analyzed in terms of, the presence of certain types of chemical bonds (e.g., C=O, C-OH, C-C, etc.) that produce characteristic absorption patterns.

inhibitory postsynaptic potential   A membrane potential across the postsynaptic membrane in a neuron that inhibits the generation of an action potential in that neuron. Inhibitory potentials are those in which the membrane is more polarized (hyperpolarized) than in the resting state.

initiation codon   An AUG CODON in messenger RNA coding for the first amino acid (methionine) in a polypeptide. All polypeptides in both prokaryotic and eukaryotic cells begin with a methionine coded for by an initiating AUG codon during the process of translation.

initiation factor (IF)   Certain proteins that catalyze the formation of the initiation complex between the mRNA and the ribosome in the process of translation.

inosine   The purine base in inosine monophosphate (IMP), the nucleotide from which the normal purine nucleotides adenosine monophosphate (AMP) and guanosine monophosphate (GMP) are synthesized biologically.

inositol   A five carbon sugar and a major constituent of the phospholipids found in cell membranes. The release of inositol in the cell membrane resulting from the action of certain growth factors and other effectors of cell differentiation

Inosine

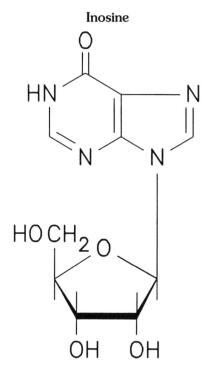

is now believed to be an important step in the way cells respond to extracellular signals.

insert In molecular biology, refers to a piece of DNA ligated into a specific site in a VECTOR for molecular cloning. The resulting recombinant molecule comprised of vector and insert is designed to allow replication of the recombinant in an appropriate host.

insertional inactivation The loss of gene activity as the result of insertion of a segment of DNA into a region critical to the expression of the gene. Insertional inactivation of genes coding for resistance to an antibiotic by insertion of a cloned DNA is often used as a means by which bacteria containing the recombinant PLASMID are selected.

insertion mutation A mutation caused by the insertion of a nucleotide or nucleotide sequence into the coding region of a gene.

in situ In the natural setting or environment; generally the intact tissue as opposed to a biochemical extract or preparation.

in situ hybridization Nucleic acid HYBRIDIZATION carried out on sections of intact tissue or chromosomes.

insulin A polypeptide hormone secreted by a part of the pancreas, known as the islets of Langerhans, which controls the entry of glucose into cells. A deficiency of insulin production is the underlying cause of diabetes.

intasome A complex between bacteriophage DNA and two proteins (Int and IHF) required in order for bacteriophage DNA to integrate into the bacterial host DNA when a BACTERIOPHAGE enters into LYSOGENY.

integral membrane protein (intrinsic protein) A protein that is integrated into the cell membrane (i.e., penetrates the membrane lipid bilayer).

integrant A cell in which a transfected gene has become stably integrated into the genome of the recipient. (See TRANSFECTION.)

integrase The enzyme that catalyzes the site-specific recombination of λ bacteriophage DNA with the bacterial host DNA, resulting in integration of the bacteriophage DNA.

integration In molecular genetics, the insertion of a foreign DNA into the genome of a recipient cell. The term is most often applied to the integration of viral DNA into the genome of an infected host (e.g., integration of the prophage in a bacterium infected by a bacteriophage).

intercalate In biochemistry, to fit a molecule in between biomolecules that are part of an array. The term is commonly applied to certain dyes that stain nucleic acids by inserting themselves in

Intasome

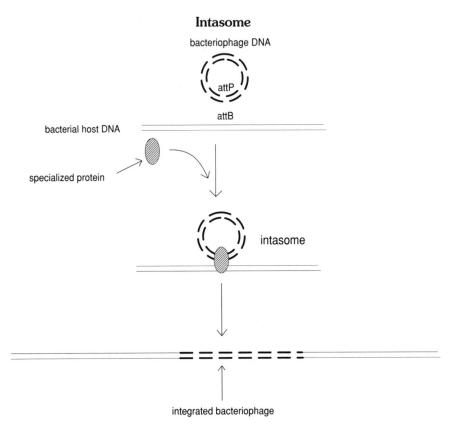

bacteriophage DNA

attP

attB

bacterial host DNA

specialized protein

intasome

integrated bacteriophage

between the purine and pyrimidine bases arrayed along the nucleic acid backbone.

interferon(s)   A group of small glyco-proteins, produced by virus-infected cells, that acts to inhibit viral infection. The interferons are divided into three types, α, β and γ, and are heterogeneous with respect to structure and mode of action. The genes for various interferons have been cloned and tested as therapeutic agents for various diseases, including Kaposi's sarcoma in HIV-infected individuals. γ Interferon has been found to induce MHC class II antigens in B cells, macrophases, and endothelial cells.

interferon, γ   See INTERFERON(s).

interleukins   Proteins, secreted by cells that interact with certain T cells as well as by the T cells themselves, that cause the T cells to proliferate. Proliferation of T cells induced by interleukins I (IL1) and 2 (IL2) are known to be involved in the T-cell proliferation response, leading to T-cell-mediated immunity.

intermediary   metabolism   That part of biochemistry dealing with how energy is derived from nutritive biomolecules and how that energy is used in the metabolism of other biomolecules.

intermediate filaments   A type of filament that makes up one kind of cytoskeleton in mammalian cells. Intermediate filaments are approximately 8–10 nm in diameter, which places them in a range intermediate between the actin and microtubule type cytoskeletal filaments. Intermediate filaments are divided into six classes: keratins, desmins, vimentins, glial filaments, neurofilaments and nuclear lamins.

interphase   The period between mitoses. (See MITOSIS.)

int 1 **gene**   An oncogene activated by the nearby integration of the mouse mammary tumor virus (MMTV), which produces mammary tumors in mice. The *int 1* gene homologue in *Drosophila melanogaster* has been shown to play a crucial role in wing development, suggesting that the mammalian *int 1* gene may play a regulatory role in development.

intracellular   Within the cell.

intramuscular   Located in, or directly administered to, muscle tissue.

intraperitoneal   Located in, or directly administered to, the cavity between the internal organs of the abdomen and the abdomen wall. Intraperitoneal inoculation of transformed cells into mice is widely used to promote the growth of transformed cells or to derive large quantities of substances they secrete (e.g., monoclonal antibodies).

intravenous   In a venous blood vessel (vein), such as the route of an intravenous injection.

intron   The DNA sequences between the EXONS of a gene.

inulin   A long polysaccharide composed largely of repeating fructose subunits. Because it is a large molecule and largely inert, inulin is used experimentally to control osmotic flow cross membranes and as a diagnostic aid for kidney function.

inversion   The alteration of cellular DNA sequences in which the orientation of a DNA segment is reversed; placed into an inverted orientation. Inversions are frequently caused by the movement of transposons, especially in cells carrying two copies of a TRANSPOSON in opposite orientation to one another.

invertase   The enzyme β-D-galactosidase that catalyzes the cleavage of lactose into the monosaccharides glucose and galactose. The enzyme is so named because its action causes the resulting sugars to undergo conversion to the opposite optical isomer (i.e., from D form to L form). Lack of invertase is responsible for lactose intolerance.

inverted terminal repeats   Segments of DNA at the ends of an insertion element, such as a transposon, that are inversions of one another, for example, AACGCTTCG and GCTTCGCAA. Inverted terminal repeats are essential for the transposability of the insertion sequence.

in vitro   Biological material outside of the normal setting (e.g., in cell or tissue culture or in cell or tissue extracts).

in vitro fertilization   Fertilization of the egg outside the body.

in vitro mutagenesis   Mutagenesis of cells in vitro (i.e., by exposure of cells in tissue culture to mutagenic agents).

in vitro packaging   The formation of the viral coat around a viral nucleic acid using biological preparations or extracts in an artificial environment (e.g., a mixture containing biological extracts, salts, necessary biological molecules, etc.).

in vitro protein synthesis   The synthesis of a polypeptide using mRNA coding for that polypeptide in an artificial mixture containing ribosomes and all the molecular components necessary to reproduce the normal process of polypeptide biosynthesis as it occurs in the intact cell.

in vitro transcription/translation   The synthesis of both mRNA and its encoded protein in an artificial mixture containing the appropriate DNA, ribosomes, and all the molecular components necessary to reproduce the normal processes of transcription and polypeptide biosynthesis as they occur in the intact cell.

in vivo In the intact cell or tissue.

iododeoxyuridine (IUdR) A synthetic nucleoside that is an inducer of Epstein–Barr virus (EBV) gene expression in EBV-infected cells that otherwise produce no viral proteins.

ion-exchange chromatography A technique for separating substances in a mixture based on passing the mixture through a column containing a matrix that binds the substances in the mixture according to their electric charge. (See CHROMATOGRAPHIC TECHNIQUES.)

ion-exchange resin A material used to separate substances in a mixture by ion-exchange chromatography. Ion-exchange resins are generally in the form of beads composed of an inert polymeric substance, such as cellulose or sepharose, covalently attached to electrically charged molecules.

ionizing radiation Any electromagnetic radiation that can knock electrons from molecules, thereby producing ions. α, β, γ radiation and X-rays are all considered to be ionizing radiation.

ionophore Any of a number of relatively small organic molecules that act to allow the osmotic passage of ions and other molecules across cell membranes that would otherwise be impermeable to them. Ionophores have been widely used experimentally to study the function of ion gradients across membranes (e.g., the Na^+–K^+ gradients in neurons).

ion-selective electrode An instrument for measuring the concentration of ions of one specific atom in a solution by measuring the current produced when a probe containing the oxidized or re-

duced form of the atoms to be tested is immersed in the solution.

isoaccepting transfer RNAs Different (tRNAs) that carry the same amino acid. (See ADAPTOR MOLECULE.)

isoantigen (alloantigen) An antigen produced by only some members of a species but not others and capable of eliciting an immune response in the individuals of the species lacking the antigen. Blood group antigens are examples.

isoelectric focusing A variation of polyacrylamide gel electrophoresis in which proteins in a mixture are separated on the basis of their individual isoelectric points.

isoelectric point The pH at which the net charge (the sum of the charges on all the individual subgroups) on a molecule is exactly zero.

isoleucine An amino acid whose side chain is

$$-CH-CH_2-CH_3$$
$$|$$
$$CH_3$$

isomerase Any of a class of enzymes that catalyzes the rearrangement of atoms in a molecule.

isoprene An organic molecule appearing in polymeric form in a number of important molecules that act as intermediates in electron transfers in various metabolic reactions in intermediary metabolism, such as ubiquinone (Coenzyme Q) and chlorophyll. Its structure is

$$CH_2=C-CH=CH_2$$
$$|$$
$$CH_3$$

Isopropyl-β-D-thiogalactopyranoside (IPTG) A synthetic analog of naturally occurring galactosides, such as lactose. IPTG is widely used in place of lactose as a potent inducer of the LAC

OPERON and, unlike lactose, is not acted upon by the enzyme β-galactosidase that it induces.

isoschizomer Two different restriction enzymes that recognize the same sequence. (See RESTRICTION ENDONUCLEASE.)

isotonic point The point at which the concentration of all solutes in a solution results in an osmotic pressure across a membrane exactly the same as the

osmotic pressure of a reference solution. This term, or synonyms, is frequently used to describe solutions that can be introduced into a biological system without causing osmotic lysis of the cells— for example, solutions injected into the bloodstream without causing hemolysis are called isotonic solutions.

isotope One of any alternative forms of an element that differ from one another in terms of the number of neutrons in the nuclei of their atoms.

J

JC virus A human virus member of the Papova group (See PAPOVA VIRUS). JC virus, closely related to the monkey virus, SV40, was first isolated from the brain tissue of a patient with progressive multifocal leukoencephalopathy (PML), a disease the virus is believed to cause.

joining gene (j gene) A DNA segment in the immunoglobulin gene cluster that joins the constant and variable immunoglobulin gene regions during B-cell maturation. During the maturation process antibody diversity is generated by

joining a constant region with numerous different variable regions.

jun An oncogene transduced by a chicken retrovirus that causes fibrosarcoma tumors. The *jun* proto-oncogene has been found to share identity with the transcription regulation factor AP-1 and apparently exerts its oncogenic effects by inducing aberrant gene transcription.

junin virus A member of the tacaribe subgroup of the arenaviruses. Junin virus, which causes hemorrhagic fever, is carried by bats and rodents.

K

kanamycin A broad-spectrum antibiotic active against both gram-positive and gram-negative bacteria and a number of types of mycoplasma. Kanamycin is an aminoglycoside derived from the soil bacterium *Streptomyces kanamyceticus*.

karyogamy The fusion of two nuclei; for example the fusion of pronuclei occurring during fertilization of the egg.

karyoplast The cell fraction containing the nucleus surrounded by a small ring of cytoplasm in cells enucleated by treatment with cytochalasin.

karyotype The characterization of the chromosomes of a cell type normally including chromosome morphology, chromosome number, chromosome banding patterns, and any abnormalities of these characteristics.

keratin   A type of intermediate filament found almost exclusively in mammalian epithelial cells and bird feathers. Mammalian keratin proteins are a large and diverse family of proteins—more than 30 polypeptides are known—of which only a small number is required for filament formation in any epithelial cell subtype.

keratinocyte   Any mammalian epithelial cell.

ketone body   Either acetone, acetoacetate, or β-hydroxybutyrate produced as the result of the accumulation of acetyl CoA due to blockage of normal glucose metabolism (e.g., occurs in diabetes).

ketose   Any sugar containing a keto (as opposed to an aldo) group:

$$\begin{array}{c} R_1 \\ | \\ C = O \leftarrow \text{keto group} \\ | \\ R_2 \end{array}$$

Khorana, Gobind (b. 1922)   An organic chemist who was among the first to develop a method for the chemical synthesis of DNA. His work in which synthetic polyribonucleotides were employed as a tool for breaking the genetic code earned him the Nobel Prize in medicine in 1968.

kilobase (kb)   A measure of the length of a nucleic acid strand equivalent to 1000 nucleotides.

kilodalton (kD)   A measure of the size of large biomolecules, but generally applied to proteins, equivalent to the molecular weight of the molecule divided by 1000. For example, a molecular weight of 125,000 corresponds to 125 kD.

kinase   A class of enzymes that catalyzes the transfer of a phosphate group from one substrate to another. Phosphorylation is a means of regulating the activities of a number of other enzymes, and thus kinases may control a wide variety of biochemical pathways through a single phosphorylation.

kinesin   A protein involved in the movement of small vesicles along a microtubule in the axons of nerves and perhaps in other cell types. Kinesin is an ATPase that uses the energy released by ATP hydrolysis to induce movement. (See ADENOSINE TRIPHOSPHATE.)

kinetochore   A dense structure in the centromeric region of a chromosome to which the spindle fibers are attached. (See CENTROMERE.)

kinetochore fibers   The microtubules extending between the kinetochore of the chromosome(s) and polar bodies that pull the chromosomes to the poles of a dividing cell during MITOSIS.

kininogen   A factor in the intrinsic blood coagulation (clotting) pathway, and one of two factors required for activation of factor XII (Hageman factor).

kinins (cytokinins)   Plant hormones that, in combination with auxins, stimulate cell division and differentation in a variety of plant tissues. Chemically, the cytokinins are purines with terpenoid side chains.

kirromycin   An antibiotic that acts by inhibiting protein synthesis on the bacterial RIBOSOME. Kiromycin forms a complex with a tRNA that prevents the elongation of the growing polypeptide chain. (See RIBOSOME, TRANSFER RNA.)

Kirsten sarcoma virus (Ki-MuSV)   A retrovirus that infects rats and produces sarcomas and erythroleukia in the infected host. Ki-MuSV carries the Ki-*ras* oncogene.

Klebsiella   An important nitrogen-fixing soil bacterium. (See NITROGEN FIXATION.)

Klenow fragment, enzyme   A subfragment of DNA polymerase I produced by proteolytic cleavage of the 103-kD enzyme by subtilisin. The Klenow fragment is the larger (68 kD) of the two subfragments produced by subtilisin treatment. This fragment retains the normal DNA polymerase and 3'→5' exonuclease activities but lacks the 5'→3' exonuclease of the intact enzyme.

Klett unit   A unit of light absorption used in measuring bacterial cell number in terms of the turbidity of bacterial liquid cultures. Klett units measured at wavelengths between 490 and 550 nm are approximately proportional to cell number during logarithmic growth.

Klinefelter's syndrome   A chromosomal aberration involving the sex chromosomes in which cells contain two X chromosomes and one Y chromosome. Afflicted individuals have the physical appearance of males but are infertile and have underdeveloped testicles and other physical abnormalities.

Kornberg, Arthur (b. 1918)   The discoverer of DNA polymerase I, which

he isolated from *E. coli* bacteria in work dating from 1956. The discovery of the first enzyme known to be responsible for synthesis of DNA won him the Nobel Prize in physiology and medicine in 1959.

Kornberg enzyme   The enzyme discovered by Arthur Kornberg: bacterial DNA polymerase I.

K-*ras*   The oncogene carried by the Kirsten sarcoma virus. K-*ras* is a member of the RAS ONCOGENE family containing Ha-*ras* and N-*ras*. The family is defined by base sequence homology of the members to one another.

Krebs cycle   See   TRICARBOXYLIC ACID CYCLE.

Krüppel gene   One of the homeobox gap genes identified in *Drosophila melanogaster*. In mutants of the Krüppel gene abdominal segments are deleted from the larva.

L

lac operon (lactose operon)   The operon containing the three genes coding for proteins involved in the metabolism of the sugar lactose and other β-galactosides: β-galactosidase, galactoside permease, and transacetylase.

lac repressor protein   A protein (produced by the *i* gene) that blocks tran-

scription of the genes in the *lac* operon by binding to the operator region of the *lac* promoter.

lactam antibiotics   A class of antibiotics whose molecular structure is derived from the lactam ring, usually penicillin and its synthetic derivatives,

The Lactose Operon (lac operon)

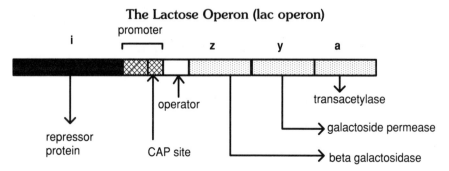

e.g., oxacillin, nafcillin, and benzyl penicillin (penicillin G).

lactamase   An enzyme, made by penicillin-resistant bacteria, that cleaves the bond between NH and C=O in the lactam ring.

$$\begin{array}{ccc} C{=}O & & COOH \\ /\ \backslash & \xrightarrow{\ lactamase\ } & / \\ (CH_2)_n{-}NH & & (CH_2)_n{-}NH_2 \end{array}$$

lactam ring   Any molecule with the general structure

$$\begin{array}{c} C{=}O \\ /\ \backslash \\ (CH_2)_n{-}NH \end{array}$$

lactate dehydrogenase   The enzyme responsible for catalyzing the conversion of pyruvate to lactic acid.

lactic acid   The product formed from pyruvate by lactate dehydrogenase when sugars are oxidized under anaerobic conditions such as occur in muscle tissue after prolonged exercise or in bacteria that thrive in low-oxygen environments.

lactic acid bacteria   Anaerobic bacteria that generate lactic acid during the process of sugar oxidation. The production of acid by lactic acid bacteria is responsible for the souring of milk and the sour taste of sauerkraut.

lactoperoxidase labeling   A technique for labeling proteins on the outside of cell membranes with radioactive isotopes of iodine (e.g., ^{125}I). Lactoperoxidase catalyzes the transfer of iodine from iodoacetamide to the tyrosine residues of the protein to be labeled.

lac Z   The β-galactosidase gene of the lactose operon.

lagging strand   During DNA synthesis, the DNA strand copied in the $5'{\rightarrow}3'$ direction. (See OKAZAKI FRAGMENTS.)

lag phase   The period between the time when a microorganism is inoculated into a nutrient broth and the time when those microorganisms enter logarithmic growth.

lamella   A thin membrane or plate dividing certain biological compartments, for example the region between the cell walls of opposing cells of certain plants (called middle lamella).

lamellipoda   A cytoplasm-containing protrusion or villus extending out of the leading edge of an animal cell during its movement along some substrate and oriented along the axis of movement.

laminar flow   A uniform, eddy-free, flow of air or liquid.

laminin   A protein component of the basement membrane that forms underneath epithelial cells where the cells adhere to a basement membrane or other substrate.

lampbrush chromosomes   Enlarged chromosomes seen in amphibian oocytes during meiotic PROPHASE. Lampbrush chromosomes are characterized by large protruding loops of transcriptionally active DNA.

lariat   An intermediate stage in the splicing out of introns during the forma-

Lariat Intermediate in RNA Splicing

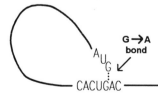

tion of mRNA in the nucleus. In a lariat, the INTRON of an mRNA precursor is cut at one end; the cut end then forms a covalent bond to a nucleotide in the interior of the intron to form the lariat structure.

laser (light amplification by stimulated emission of radiation)   Laser light is created by causing a group of atoms to emit photons in synchrony. Lasers are used in various types of molecular biological analyses (e.g., flow cytometry).

Lassa fever virus   An RNA-containing virus member of the arenavirus family. First discovered in Nigeria, the virus is known to cause an acute infection characterized by fever, malaise, throat lesions, and pneumonia.

late genes   In viral infection, a set of genes always expressed late in the life cycle of the virus. Generally, the late genes code for proteins required for packaging of the viral DNA, which is replicated early in the viral life cycle.

lateral meristem   Meristem tissue lining the plant stem. Because the meristem is mitotically (growth) active, the lateral meristem is responsible for growth in diameter of the plant stem.

LD$_{50}$ The dose of a test drug fatal to 50% of test animals to which it is administered.

leader peptidase   An integral membrane protein that catalyzes the cleavage of the leader sequence during the insertion of preproteins into their target membranes.

leader sequence   A short segment at the beginning of a newly synthesized peptide serving as a signal that the peptide is to be transported outside the cell —that is, secreted or deposited on the outer surface of the cell membrane.

leaky mutant   A mutant microorganism in which the normal properties continue to be expressed at a low level or in which the mutation is only partly expressed.

lecithin   The common name for the membrane phospholipid, phosphatidyl choline. Lecithin is believed by some to possess detergent properties capable of dissolving cholesterol present in arterial plaques.

lectins   Plant-derived proteins that bind to specific polysaccharides. Lectins are used to label the cell surfaces of, or to agglutinate, cell types bearing the particular polysaccharides.

Lederberg, Joshua   (b. 1925) One of the discoverers of the phenomenon of transduction by bacteriophage and gene transfer between bacteria by conjugation. Lederberg was awarded the Nobel Prize in 1958.

legume   A member of the pea family of plants. The roots of leguminous plants maintain a symbiotic relationship with nitrogen-fixing bacteria so that the legumes are a rich source of nitrogen stored in the form of nitrates.

lentivirus   Literally, "slow (=lentus) virus." A type of retrovirus producing a chronic, generally subclinical infection, such as visna virus, which infects the brain cells of sheep. However, infection may invoke an immune response that can result in demyelination of the nerve cells. It is believed that some demyelinating diseases in humans may follow the same paradigm.

Lesch–Nyhan syndrome   A genetic disease characterized by mental retardation and loss of coordination. The disease, manifested by the age of 2, is due to the lack of the enzyme hypoxanthine-guanine phophoribosyltransferase (HGPRT) and was identified by Michael Lesch and William Nyhan in 1964.

lethal locus   A genetic locus where mutations tend to prove lethal to organisms carrying the mutation(s).

lethal mutation   Any mutation whose presence leads to the death of the organism in which the mutation is present.

leucine   The amino acid that contains an isobutyl group as a side chain:

$$CH_3 \diagdown CH-CH_2-$$
$$CH_3 \diagup$$

(See ISOLEUCINE.)

leukemia   A cancer of the blood characterized by the uncontrolled proliferation of white blood cells (LEUKOCYTES).

leukocyte   A white blood cell. The white blood cells largely comprise cells of the immune system.

levorotatory isomer   One of the two main classes of optical isomers. When polarized light is passed through a solution of a levorotatory isomer, the plane of polarized light is rotated in a counterclockwise direction (i.e., to the left) from the point of view of the observer.

library   A large set of cloned DNA sequences from some specified source, e.g., cDNAs or fragments of chromosomal DNA derived from a certain tissue or cell type. (See COMPLEMENTARY DNA.)

ligand   Any molecule bound by a specific receptor for that molecule.

ligand-gated channels   Channels in cell membranes that permit the passage of ions through the membrane when, and only when, a specific ligand is bound to its membrane receptor (ligand gating). Ligand-gated channels are the means by which nerve impulses are propagated when a neurotransmitter produced by one neuron binds to a receptor on the membrane of another neuron.

ligase   A type of enzyme that catalyzes the formation of a convalent bond between the $5'PO_4$ end of one strand of nucleic acids and the $3'OH$ end of another.

ligase, DNA   Catalyzes the linkage between a 5'-terminal phosphate group at the end of one double-stranded DNA and a 3'-terminal hydroxyl group at the end of another.

ligation   The chemical linking of the free ends of two nucleic acids to form one larger strand out of two smaller ones.

light chains (κ, λ)   Two short Peptides in IgG antibody molecules. The λ type is distinguished from κ type on the basis of antiserum to light-chain proteins; immunoreactivity of the light chains show they are either κ or λ type, but not both. κ and λ light chains are secreted by myeloma tumors and appear in the urine of myeloma patients, where they were originally called Bence-Jones proteins. This discovery helped to elucidate the structure of the IgG molecule.

light-dependent reactions   Those chemical reactions in photosynthesis that require light. The so-called light reaction(s) involve the capturing of light energy by pigments including chlorophyll in the form of high-energy electrons.

light-independent reactions   Chemical reactions in photosynthesis carried out in the absence of light. The so-called dark reactions are responsible for trapping carbon dioxide and water to create sugars.

lignin   A phenolic polymer forming a matrix in which the cellulose fibers of the plant cell wall are embedded. Lignin forms the "cement" that holds the fibers in place and also lends tensile strength to the cell wall.

limit digest   The product of a degradative enzymatic reaction in which the substrate has been digested to the maximal extent possible. Limitation of the extent of digestion may be imposed by physical constraints, such as clumping of the substrate material or failure of the substrate to completely enter solution.

lincomycin (lincocin)   An antibiotic produced by *Streptomyces lincolnensis*. The antimicrobial action of lincocin is due to its binding to the large subunit of the RIBOSOME that prevents synthesis of peptides by blocking elongation of partially synthesized peptide chains.

Lineweaver–Burk plot   A plot of the reciprocal of substrate concentration $(1/[S])$ versus the reciprocal of the reaction velocity $(1/V_0)$ for an enzyme-catalyzed reaction; also called a double reciprocal plot. This type of plot is useful for graphical determination of K_m and V_{max} for a given enzymatic reaction.

linkage   The degree to which any two genetic markers are associated with each other as determined by the frequency with which the two markers appear together in the same individual during genetic transmission, such as in offspring or in microorganisms in which the genetic markers have been transferred by transduction or conjugation, and so on. Genetic linkage is related to, but is not the same as, the physical distance between the two markers.

linkage map   A genetic map based on genetic linkage as opposed to actual physical distances.

linked genes   Genes located within close enough physical proximity to one another that they appear together in virtually all organisms to which one or the other is transmitted.

linker   A synthetic molecule serving as a molecular bridge between two other molecules, such as a synthetic oligonucleotide that joins two DNA fragments.

linker scanner mutations   The replacement of a segment of DNA with a synthetic oligonucleotide (a linker) containing mutations whose effect on the activity of a PROMOTER is to be tested. The promoter sequence altered by the linker is tested for activity *in vitro* (e.g., by CAT assay). (See CHLORAMPHENICAL ACETYL TRANSFERASE ASSAY.)

linking number (L)   A measure of supercoiling that represents the total number of times a twisted double-stranded DNA crosses itself.

linking number paradox   In experimental determinations of the number of winds of DNA around each nucleosome, the linking number paradox refers to the discrepancy between the values obtained by different experimental methods. For example, digestion of NUCLEOSOMES by ENDONUCLEASES gives about 1.8 DNA coils per nucleosome, but measurements based on supercoiling of DNA after the nucleosomes are removed give a value of about 1 coil per nucleosome. The discrepancy is of theoretical significance for models of the helical structure of the DNA molecule.

lipase   Any of a variety of enzymes that catalyzes the breakage of an ester linkage in a lipid, thereby participating in the breakdown of that lipid.

lipid   Any of a variety of oily, highly insoluble biomolecules associated with cell membranes and fatty tissues. The six major classes are fatty acids, triglycerides, phosphatides, sphingosines, waxes, and cholesterol derivatives.

lipid bilayer   A thin film of regular thickness that, under certain conditions, is spontaneously formed by amphipathic lipids when they are placed in water. The plasma membranes of animal cells

Formation of the Lipid Bilayer

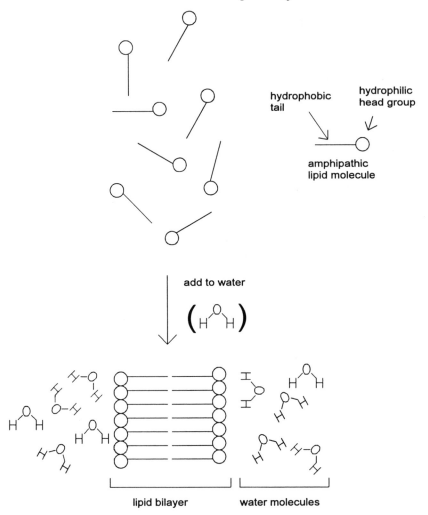

lipid bilayer   water molecules

is formed, in large part, from naturally occurring bilayers of phospholipids.

lipopolysaccharide   Lipids bound to polysaccharides.   Lipopolysaccharides are found attached to the outsides of many cell membranes, including bacterial cell membranes.

lipoprotein   Complexes of lipids and proteins. The most important and abundant examples are lipoproteins that transport fats in the blood and are classi-

fied mainly in terms of their densities: high-density lipoproteins (HDLs), intermediate-density lipoproteins (IDLs), low-density lipoproteins (LDLs), and very low density lipoproteins (VLDLs).

liposome   A synthetic structure comprising a lipid bilayer completely enclosing an interior cavity in which may be carried various substances of interest. Because the lipid bilayer of the liposome can spontaneously fuse with the lipids of the cell plasma membrane, liposomes

Packaging of DNA by Liposomes

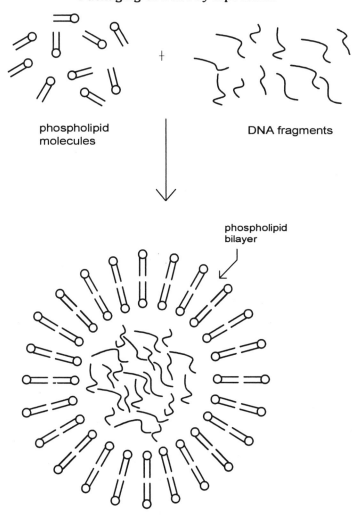

phospholipid
molecules

+

DNA fragments

phospholipid
bilayer

liposome

are being studied as vehicles to specifically introduce drugs or other bioactive chemicals directly into target cells.

lipotropic agents   Lipid solvents or lipid degrading enzymes as virus inactivating agents.

liquid chromatography   A term for any chromatographic technique in which samples are dissolved or suspended in a liquid medium and subjected to chromatographic separation by passing the liquid phase over a solid matrix. Liquid chromatography is a general term that may encompass a wide variety of specific chromatographic principles, including ion-exchange or adsorption, size-exclusion, or reverse-phase chromatography. (See CHROMATOGRAPHIC TECHNIQUES.)

locus   The position occupied by a gene or a genetic marker on a chromosome.

LOD score   A numerical value used to indicate genetic linkage between two markers, defined as LOD = $\log_{10}(P_{linked}/P_{unlinked})$, where P_{linked} is the probability that the frequency with which two markers segregate from one another in the offspring of a mating could have occurred if the markers were linked; $P_{unlinked}$ is the probability that the frequency with which two markers segregate from one anther in the offspring of a mating could have occurred if the markers were unlinked from one another. By convention, a LOD score of 3 is considered the threshold value for declaring that two markers are linked.

logarithmic (growth) phase   Term describing growth of a culture of microorganisms under conditions where, on average, one organism gives rise to two daughters at a consistent uniform rate. Logarithmic phase of growth follows lag phase during which some organisms may not reproduce or give rise to only one daughter during reproduction.

long-period interspersed sequences (LINES)   Moderately repeated sequences interspersed throughout the mammalian genome. LINES are believed to represent retroposons that have inserted themselves in many locations throughout the genome.

long terminal repeat (LTR)   Specialized sequences located at the 5'- and 3'-termini of the genome of retroviruses. LTRs mediate the integration of the RETROVIRUS into the host genome and regulate the transcription of the retrovirus genes. Because many LTRs contain strong enhancer elements, they are often used in synthetic constructs to express foreign or engineered genes in recipient cells.

loop, looped domains   A single-stranded region in either RNA or single-stranded DNA that forms a hairpin structure. The sequence between inverted repeating sequences forms looped domains because of base pairing between the inverted repeating sequences.

low-density lipoprotein (LDL)   A class of lipoprotein particles carrying cholesterol esters to cells with specialized LDL receptors. Receptor-bound LDLs are taken up into the cell where the cholesterols are metabolized.

low-density lipoprotein (LDL) receptor   A transmembrane protein in liver cells that binds specifically to LDLs after which the LDLs are taken into the cells by ENDOCYTOSIS and degraded. Uptake of LDLs by this mechanism is one of the major pathways for the metabolism of LDL-associated cholesterol.

L-phase variants   Bacterial variants lacking a cell wall. L-phase variants are produced under conditions in which cell-wall synthesis is inhibited (e.g., in the presence of penicillin). The L-phase variants pass through filters that retain normal bacteria.

luciferase   An enzyme isolated from fireflies and responsible for their characteristic light flashes. Luciferase catalyzes the decarboxylation of the substrate luciferyl adenylate to generate light. This reaction has been exploited as a nonradioactive means of labeling molecules for analytical purposes.

$$\text{luciferin} \xrightarrow{\text{ATP}} \begin{array}{l}\text{luciferyl adenylate}\\ + \text{AMP}\end{array}$$

$$\xrightarrow[\text{oxygen}]{\text{luciferase}} \begin{array}{l}\text{oxyluciferin} + CO_2\\ + \text{AMP} + \text{light}\end{array}$$

Luria, Salvador   (b. 1912) A geneticist, winner of the Nobel Prize in medicine in 1969, whose work with bacterial and bacteriophage mutants led to the elucidation of gene structure.

luteinizing hormone (LH)   A glycoprotein hormone released from the anterior pituitary. LH stimulates oocyte maturation and ovulation and progesterone secretion in the ovary.

luxury genes A vernacular term to describe nonessential genes for basic cell functions but are made by only a certain cell type and perform a function necessary to the functioning of the organism as a whole (e.g., hemoglobin genes). (See DIFFERENTIATION ANTIGEN.)

lympho adenopathy virus (LAV) Another name for HIV, the virus that causes AIDS. The designation LAV was used by the French discoverers of the virus at the Pasteur Institute.

lymphocyte The subclass of white blood cells responsible for carrying out the immune response. Lymphocytes are further subdivided into T- and B-lymphocytes and are found mostly in the thymus, lymph nodes, spleen, and appendix.

lymphokines (interleukins) Hormones secreted by certain antigen-processing cells of the immune system that cause T cells specific for the antigen to proliferate.

Lyon effect The silencing (prevention) of expression of the genes on one of the two X chromosomes in the cells in a female animal. Silencing of the X chromosome occurs at random and is maintained in all the progeny cells.

lyophilization The removal of water from a frozen biological specimen by placing the specimen in a vacuum, often referred to as "freeze drying."

lysate The resultant mixture of cell debris and soluble cytoplasmic substances resulting from mass cellular lysis of a cell culture or tissue.

lyse (lysis) A breaking open of cells by damage to the cell membrane by any mechanical, biological, or chemical agent.

lysine An amino acid with an amino butyl side chain $-(CH_2)_4-NH_2$. The amino group makes lysine a basic amino acid.

lysogen A bacterial strain harboring a lysogenic virus (bacteriophage). (See PROPHAGE.)

lysogenic The property of certain bacteriophage mutants to integrate into, and remain dormant in, the bacterial host DNA. Although lysogenic bacteriophage do not immediately cause lysis, they may be induced to do so by various chemical agents or ultraviolet light.

lysogeny The state of being lysogenic.

lysosome A cytoplasmic organelle in eukaryotic cells in which digestion of particulate material brought into the cell via endocytosis or phagocytosis occurs. Lysosomes are characterized by a highly acidic internal environment and the presence of enzymes (acid hydrolases) that carry out the digestive process.

lysozyme An enzyme deriving its name from its ability to cause certain bacteria to lyse. Lysozyme acts by cleaving polysaccharides in the bacterial cell wall. Lysozyme is used as a reagent in preparing DNA from bacteria and in the creation of protoplasts (plant cells or bacterial cells from which the cell wall has been removed).

lytic cycle The events in the growth cycle of a lytic virus.

lytic virus Any virus that, as part of its life cycle, causes lysis of the host cells it infects.

M

macrolides A group of antibiotics with a large aliphatic ring structure with many hydroxyl and keto groups. Erythromycin is the best-known member of this group of antibiotics.

macrophage A large phagocytic white blood cell. Macrophages travel in the blood but are capable of leaving the blood to enter tissue. They function as a defense by ingesting invading bacteria and other foreign cells as well as by removing particulate debris.

magic spot nucleotides Nucleotides with two or more phosphate groups on both the 3'- and 5'-carbon atoms that accumulate in bacterial cells during the STRINGENT RESPONSE.

main band The band corresponding to the bulk of DNA, as opposed to satellite DNA, when a preparation of mammalian genomic DNA is subjected to density gradient centrifugation analysis.

major histocompatibility complex (MHC) A cluster of genes present in the genomes of most higher vertebrates that code for cell surface proteins that are recognized as the main transplantation antigens when cells are transplanted to a foreign environment. The MHC proteins are therefore the antigens mainly responsible for provoking graft rejection in animals receiving foreign tissue grafts. (See TRANSPLANTATION ANTIGENS.)

malaria A chronic disease characterized by periodic acute attacks of chills and fever. The disease is the result of infection by the sporozooite parasite *plasmodium,* which lives in red blood cells and is transmitted to humans via the *Anopheles* mosquito.

maltase The enzyme that catalyzes the breakdown of the malt sugar maltose:

$$maltose \xrightarrow{maltose} 2 \ glucose$$

maltose binding protein A protein produced by the bacterium *Eschericia coli* used to transport the disaccharide sugar maltose across the bacterial plasma membrane.

maltose binding protein (MBP) vector An expression vector designed to facilitate the purification of the proteins expressed via the VECTOR. In MBP vectors the gene to be expressed is fused to the gene coding for the MBP. The fusion protein expressed by the vector can be purified by running a cell extract over an amylose column.

mammalian cell culture The maintenance of mammalian cells outside the body using synthetic media to meet the nutritional requirements normally supplied by the blood.

mammalian expression systems The term for expression vectors specifically designed to express cloned genes in mammalian cells. (See EXPRESSION SYSTEM, VECTOR.)

mannose A sugar and an optical isomer of the main energy producing sugar glucose. Because mannose differs from glucose at only one of the six carbons, it is called an epimer of glucose and can be converted directly into glucose by enzymes.

map distance A means of defining distance between two markers on a chromosome; measured in terms of centimorgans (cM).

Marburg virus A virus discovered as a contaminant of tissues from African green monkeys in Marburg and Frankfurt, Germany, in 1967. The classification of Marburg virus is unclear. The disease is characterized by fever, rash, gastrointestinal upset, and central nervous system involvement and is potentially fatal.

marker Any genetic element that produces a variation in expression of a trait (e.g., hair color) and resides at a particular locus.

mass spectrometry An analytical technique for determining the molecular structure of an unknown compound by observing the paths taken by molecule fragments when forced to migrate in a magnetic field.

mast cell A connective tissue cell located near capillaries and most abundant in the lung, skin, and gastrointestinal tract. Mast cells possess receptors for IgE and release histamine when bound to IgE. The release of histamine is responsible for the runny nose, itchiness, and other respiratory symptoms of allergy.

master regulatory genes A cluster of genes governing the development of the major structural features during embryogenesis—for example, the bithorax complex in *Drosophila melanogaster* is responsible for development of the abdominal and thoracic segments.

mating-type locus (MAT) A locus containing master regulatory genes in yeast that determines the male and female mating types, designated as α and a.

matrix attachment regions (MAR) (scaffold attachment regions, SAR) Specific DNA sequences at which attachment to the NUCLEAR SCAFFOLD network occurs.

maturing face [trans face (Golgi)] The outermost membrane in a Golgi stack. The maturing face of the Golgi stack is the place where proteins processed in the Golgi exit for various cellular destinations. (See GOLGI APPARATUS.)

Maxam–Gilbert sequencing A technique for determining the sequence of a nucleic acid by chemical treatments that cleave the nucleic acid strand at only one of the four nucleotide bases (i.e., adenine, cytosine, guanine, thymine, or uracil), depending on the chemicals used. The fragments produced by chemical cleavage are then separated by electrophoresis, and the sequence is determined from the size of the different fragments.

McClintock, Barbara (1902–1992) A geneticist whose work on transposable genetic elements in corn won her the Nobel Prize in medicine in 1983. (See TRANSPOSON.)

mDNA That portion of the genomic DNA that hybridizes to all the messenger RNAs (mRNA) produced in all tissues. It is presumed to represent the sum total of all genes whose expression is required for processes required in all cell types.

medium The nutrient broth used to grow cultures of cells, bacteria or microorganisms.

meiosis Cell division that takes place in the reproductive tissue and produces gametes (sperm and egg in animals). The meiotic process leaves each daughter cell, which becomes a gamete, with half the number of chromosomes as found in other cell types in the body.

melanin The brown, reddish, or black pigment that, in mammals, gives skin and hair their characteristic color(s). Melanin is derived from the amino acid tyrosine and is also found in parts of the brain and eye where its function is unknown.

melanocyte A specialized cell type, found beneath the epidermal layer of skin, that produces melanin for the purpose of skin pigmentation. Melanin made in the melanocyte is passed on to the upper layers of skin through dendritic cell processes.

melanoma   A highly malignant skin cancer involving melanocytes.

melting, of DNA   The breakage, by heating, of the hydrogen bonds that hold the double-stranded helical structure of DNA together. Upon melting, DNA changes from double stranded to single stranded.

melting temperature   The temperature at which 50% of double-stranded DNA is turned into single-stranded DNA.

membrane   In general, a flexible sheet or layered material that separates two chemically different environments. Biologically, membranes are made up primarily of lipids and proteins and are the structures defining the compartments of the cell, organelle, and nucleus. Synthetic membranes are used biochemically to separate chemically different liquids and gases from one another for experimental purposes.

membrane potential   The electric potential (voltage) created by the difference in ion concentration on different sides of a membrane. Membrane potentials drive certain kinds of transport systems through the membrane and are responsible for nerve impulses. (See PROTON GRADIENT.)

membrane ruffling   A wavelike movement observed at the leading edge of a cell membrane during movement; the location of the ruffled portion of the membrane indicates the direction of cell movement.

memory, immunologic   The ability of the immune system to respond to antigens to which it has previously been exposed; the maintenance of immunity to an antigen over long periods of time.

memory cells   Small populations of B and T cells of the immune system that produce antibodies or have receptors for an antigen and appear after an initial exposure of the organism to that antigen.

Mendel, Gregor   (1822–1884) The founder of the field of genetics. His experiments on crossbreeding of peas led to the first formulation of the principles of heredity.

Mendelian genetics   Genetics based on the concept completely independent, randomly reassorting genetic elements.

Mendel's law   The principle that genetic elements controlling individual traits can reassort themselves independently of one another during the reproductive process. The postulated genetic elements were later found to be based on physically discernible structures, chromosomes.

mercaptoethanol   A widely used reducing agent, particularly useful in biochemical procedures where breakage of disulfide bonds is desirable:

$$-S-S-\xrightarrow{\text{mercaptoethanol}} -SH \quad HS-$$

6-mercaptopurine   A purine analog whose DNA-damaging effects particularly target the production of T-lymphocytes. Thus, 6-mercaptopurine is given as an immunosuppressant drug to prevent graft rejection.

meristem   The mitotically active tissue in higher plants that, through cell division, forms new plant tissues. Meristematic cells are found in the root (apical meristem) and along the outside of the stem (lateral meristem).

merozygote   The stage in bacterial conjugation in which the recipient bacterium, prior to division, contains two bacterial chromosomes.

mesophile   Bacteria that grow in the narrow temperature range of the mammalian body, from about 37°C to 44°C.

messenger RNA (mRNA)   A ribonucleic acid strand carrying the genetic code for a protein. The mRNA is copied from DNA in the nucleus, processed, and transported to the cytoplasm, where its code is read on RIBOSOMES and translated into a polypeptide.

metabolic disease   A disease stemming from a defect in an essential metabolic pathway.

metabolic pathway   A series of successive biochemical steps in the metabolism of a nutrient molecule in which the input molecule is progressively altered until a specific final metabolite is produced.

metabolism   The process of altering a nutrient molecule via a metabolic pathway for energy production or creation of important biomolecules (e.g., amino acids, hormones, nucleic acids, etc.).

metabolite   One of the intermediate molecules generated during the metabolism of a nutrient.

metalloenzyme   An enzyme requiring a metal atom(s) for normal activity.

metallothionein   Any of a class of metal-binding proteins that play a role in preventing toxicity due to metal accumulation in cells. Because metallothionein synthesis is induced by the presence of metal ions, the metallothionein PROMOTER has been widely used to control the expression of genetically engineered genes.

metamorphosis   The maturational process in amphibians as exemplified by the transition from tadpole to adult frog.

metaphase   The phase in MITOSIS in which the chromosome pairs are aligned along the axis of the cell just prior to telophase.

methane   The simplest hydrocarbon made up of one carbon and four hydrogen atoms (CH_4); a gas generated from carbon dioxide by certain bacteria during the oxidation of fatty acids.

methanogenic bacteria   Bacteria that generate methane gas during metabolism.

methanol   The alcohol of methane (CH_3-OH), also known as wood alcohol.

methanophile   (methanotroph) Any of a class of bacteria that derives its energy from the metabolism of methane.

methicillin   A synthetic derivative of penicillin created by the addition of a dimethoxyphenyl group to the side chain of penicillin. Since methicillin is not susceptible to the action of penicillinase, it can be used in cases of infection by penicillin-resistant bacteria.

methionine   One of the two sulfur-containing amino acids whose side chain is $-CH_2-CH_2-S-CH_3$. Methionine is an essential amino acid and is important as a methyl group donor.

methionine-enkephalin   (met-en-kephalin)   A short peptide with structure; Try-Gly-Gly-Phe-Met. Met-enkephalin is one of a class of pain-inhibiting neuropeptides, known as endorphins, that act by binding to the opioid receptor in the brain.

methyl-accepting chemotaxis proteins (MCPs)   A class of bacterial transmembrane proteins involved in CHEMOTAXIS. The portion of the MCPs extending into the bacterial cytosol becomes methylated when the portion of the protein extending outside the cell binds an attractant substance, but becomes demethylated if a repellent substance is bound.

methylation, of nucleic acids   The addition of methyl groups to nitrogen atoms of bases in nucleic acids. Methylation of nucleic acids is known to serve at least three functions: (1) Methylation of the bases in DNA is believed to be a mechanism for controlling gene expression (methylated DNA is not expressed). (2) Methylation of the 5′-terminal guanine in mRNA is required for the mRNA to be functional. (3) Methylation of restriction enzyme sites in the DNA of bacterial cells that make restriction enzymes as a defense against invading bacteriophage; methylation of the sites on the bacterial DNA prevents cleavage of the DNA by its own restriction enzymes.

methylcellulose   An inert polymeric substance used to increase the density of culture medium in order to maintain growing microorganisms in suspension.

5-methylcytosine (5MeC)   A modified form of cytosine to which a methyl group has been added by a methylase. These modified residues are found at specific sites along the DNA and provide hotspots for transition-type mutations. They are readily spontaneously deaminated, resulting in the conversion of 5MeC to thymine, leaving a mispaired G-T base pair. Upon subsequent DNA replication, one newly synthesized strand will contain the A-T mutation.

methyl tetrahydrofolate   A form of the B vitamin folic acid that acts as a coenzyme in methyl group-transferring reactions in the synthesis of purines.

methyl transferases   A set of enzymes that catalyzes the transfer of methyl groups from S-adenosylmethionine (SAM) to another substrate, particularly nucleic acids or nucleic acid precursors.

micelle   A more or less spherical structure spontaneously assumed by amphipathic lipids when mixed with water. The polar portion of the lipid is oriented outward, in contact with the water molecules, and the hydrophobic portions of the lipids are in the interior of the spheroid.

Michaelis–Menten constant (K_M)   A reaction rate constant pertaining to enzyme-catalyzed reactions:

$$\text{E+S} \underset{k_{-1}}{\overset{k_1}{\leftrightarrow}} \text{ES} \overset{k_2}{\leftrightarrow} \text{E+P}$$

where E=free enzyme, S=substrate, ES=enzyme–substrate complex, P=product of the reaction, and k_1, k_{-1}, k_2=reaction rate constant for the individual reactions. K_M is defined by $K_M = (k_2+k_{-1})/k_1$.

Michaelis–Menten equation   Defines the relationship between K_M, the reaction rate velocity (v), the maximum reaction rate attainable at a given concentration of enzyme, and the concentration of substrate [S]: $v=v_{max}[S]/([S]+K_M)$.

microaerophile   A microorganism that is neither aerobic nor anaerobic but can grow under conditions of very limited oxygen.

microbe   A microorganism. Often used as a synonym for germ.

microcapsule   A very thin version of the capsule that covers the bacterial cell wall; a gel-like, largely polysaccharide matrix that protects the bacterium from PHAGOCYTOSIS.

micrococcal nuclease   An ENDONUCLEASE, isolated from *Staphylococcus aureus,* that cleaves DNA strands by breaking the deoxyribose-phosphate backbone of DNA at the 5′-carbon atom. Micrococcal nuclease is sometimes used in place of DNase I for mapping protein binding sites on DNA. (See FOOTPRINT.)

microfibrils   A bundle of fine cellulose fibers composing the plant cell wall.

microfilament   An actin-containing filament that makes up one type of CYTOSKELETON in mammalian cells. Microfilament movements are believed to be the basis for cells in culture.

microglobulin   A short peptide noncovalently bound to the class I MAJOR HISTOCOMPATIBILITY COMPLEX glycoprotein.

microgram   0.000001   or 10^{-6} grams.

Microsomes

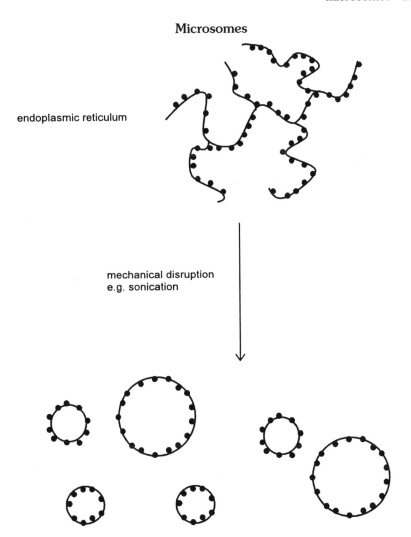

endoplasmic reticulum

mechanical disruption
e.g. sonication

resealed ER fragments
(microsomes)

microheterogeneity Slight variation in the nucleotide sequences of a repeated unit of DNA. For example, the spacer regions in the HISTONE genes are copies of one another, but there is some slight variation in the RESTRICTION FRAGMENT profiles, which is referred to as microheterogeneity.

microinjection The injection of materials directly into individual cells using a small glass micropipet.

micromanipulator An instrument for guiding extremely small instruments (e.g., microinjection pipets and microelectrodes) into individual target cells under a microscope.

micron (μm) 0.000001 or 10^{-6} meter.

microsomes An experimental preparation derived by fragmentation of the ENDOPLASMIC RETICULUM (ER); the small, roughly spherical bodies consisting of bits of ER membrane that form spontaneously when animal cells are broken. Microsomes are categorized as rough or smooth, depending on whether they are derived from rough or smooth ER. They are experimentally significant because, unlike the intact ER, they are not difficult to isolate and thus provide a convenient means of studying ER function.

microspikes   Very thin (0.1 μm diameter $\times$ 5–10 μm long), actin-containing projections that protrude from the membrane of cultured animal cells.

microtiter agglutination test   A microscale test for the presence of an antibody or an antigen based on the presence of a precipitate formed when an antigen and its corresponding antibody bind to one another. The test is carried out in a series of small wells in a plastic tray (microtiter tray) in which a fixed amount of antigen (or antibody) is added to wells containing dilutions of a sample of an unknown amount of the corresponding antibody (or antigen). Semiquantitative results are obtained as the highest dilution at which a precipitate can no longer be detected by eye.

microtome   An instrument for creating extremely thin slices (sections) of biological specimens for miocroscopic examination.

microtubule   Small tubules, comprised of a protein called tubulin, attached to the centromeres of chromosomes and responsible for the segregating movement of chromosomes during MITOSIS.

microtubule-associated proteins (MAPS)   A large class of proteins believed to stabilize microtubules by binding to their tubulin subunits. Because MAPS can bind to several tubulin molecules at once, they accelerate the rate at which microtubules are polymerized

from the tubulin subunits. MAPs are also believed to serve as a binding material that glues microtubules to other proteins and cellular structures such as the chromosome CENTROMERE.

microtubule organizing center   See CENTROSOME.

microvillus   A fingerlike projection that is actually an actin-filament-filled outpocketing of the cell membrane. Because microvilli are especially abundant on absorptive cells such as intestinal epithelial cells, they are thought to function as a mechanism for increasing the absorptive surface area of the cell membrane.

migration-inhibitory factor (MIF)   A factor(s) produced by certain T cells after stimulation by an antigen that inhibits the chemotactic response in MACROPHAGES. (See CHEMOTAXIS.)

mil   The oncogene of the chicken sarcoma virus, believed to function as a serine kinase.

milk agent   Mouse mammary tumor virus (MMTV), a RETROVIRUS causing cancers in mice. It was first identified as the agent that causes cancers in suckling mice nursed by mothers carrying the virus and was among the first tumor-causing viruses to be described.

milligram   0.001 or 10^{-3} gram.

Milstein, Cesar   (b. 1927) The researcher who, together with Georges Kohler, developed the technique of creating HYBRIDOMAS for the production of monoclonal antibodies by fusion of mouse spleen lymphocytes with myeloma cells. This work won him the Nobel Prize in medicine in 1984.

minicells   A daughter cell produced by cell division of a certain type of bacterial mutant lacking a chromosome. Since minicells lack the bacterial chromosome, those containing DNA have been used to study aberrations of DNA replication and the properties of nonchromosomal DNAs such as PLASMIDS.

minichromosome   The NUCLEO-SOME-bound form of polyoma or SV40 DNA found in the nuclei of the virus-infected cells.

minimal medium   A bacterial medium containing inorganic salts, inorganic nitrogen, and a simple sugar; the minimal requirements necessary to support bacterial growth. Minimal medium was classically used for the detection of mutants unable to synthesize an essential biochemical (e.g., an amino acid or nucleoside). (See DEFINED MEDIUM.)

mismatch repair   A type of excision repair process that targets any region of DNA in which damage or mutation has resulted in a region where nucleotide bases on complementary DNA strands are improperly paired with one another. In the bacterium *Eschericia coli,* mismatch repair involves the actions of the genes *mutH, mutL, mutS, and mutU.* (See EXCISION REPAIR.)

missense mutation   A point mutation that results in the replacement of one amino acid with another in the protein coded for by the gene in which the mutation occurs.

mitochondrion   A cytoplasmic organelle responsible for the bulk of energy production in eukaryotic cells. The mitochondrion is the site at which the electron transport process and the Krebs cycle portions of sugar metabolism take place.

mitogen   Any agent, such as a growth factor, that stimulates a cell to divide.

mitomycin C   One of a class of anti-tumor antibiotics (the mitomycins) isolated from the soil bacterium *Streptomyces caispitosus.* Mitomycin C exerts its antibiotic effects as an inhibitor of DNA synthesis.

mitosis   The process of cell division in a non-gamete-producing cell. Mitosis differs from meiosis because the diploid number of chromosomes is maintained in the daughter cells.

mitotic apparatus   A term used to describe the mitotic spindle apparatus that consists of microtubule bundles attached at one end to the CENTROMERE of the chromosome and at the other to the CENTRIOLE located at one of the two cell poles.

mitotic index   The percentage of cells in mitosis at any given moment.

mitotic recombination   Crossing over of chromosomal segments between homologous chromosomes in a somatic cell. Such recombination is a normal event in MEIOSIS but occurs rarely in SOMATIC CELLS; the recombined chromosomes are not passed on to progeny.

mitotic shake-off   A method for obtaining a cell-cycle-synchronized population of cells in tissue culture. The method depends on the fact that cells engaged in mitosis are not well attached to the bottom of the tissue culture vessel; therefore the subpopulation of cells easily detached by light shaking are, for the most part, in mitosis. (See SYNCHRONOUS CULTURE.)

mitotic spindle   The microtubule part of the mitotic apparatus.

MN blood group   A group of red blood cell surface glycoproteins (oligosaccharide derivatives of the protein glycophorin) that form a blood group family, distinguishable on the basis of naturally occurring antibodies, distinct from the ABO blood group.

mobile genetic element(s)   Insertional elements (IS).-See TRANSPOSONS.

The Mitotic Spindle

polar regions

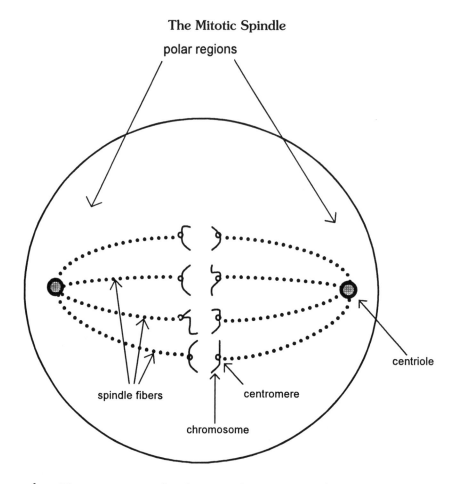

centriole

spindle fibers

centromere

chromosome

molar The concentration of a solution given as moles of the solute per liter of solution.

mold The filamentous, multicellular subfamily of the fungi.

mole A measure of the amount of a particular molecule such that 1 mole = 6.023×10^{23} molecules (Avogadro's number). One mole of a substance has a weight, in grams, equal to its molecular weight.

molecular evolution The field of study devoted to establishing evolutionary relationships between species by analysis of the relatedness (homology) of the sequences of the nucleic acids or proteins of different organisms.

molecular genetics The study of the molecular basis of genetics; the structure and function of the DNA sequences in genes and control of gene transcription and expression.

molecular weight A measure of the mass of a molecule based on a system where the mass of the hydrogen atom is taken as 1 and all other atoms are then assigned a molecular weight relative to hydrogen. (See DALTON.)

molecule Any group of covalently bonded atoms.

molt   In arthropods, a shedding of the outer covering (the exoskeleton) during maturation to accommodate body growth.

monocistronic RNA   A bacterial mRNA that codes for a single protein.

monoclonal antibody   An antibody produced by the daughter cells derived from a single antibody-producing cell.

monocotyledon   A subclass of the higher plants known as angiosperms characterized by a single, as opposed to a double, seed leaf.

monocyte   A large leukocyte with phagocytic properties. It is distinguished from other PHAGOCYTES by its size and small cytoplasmic granules.

monomer   The single subunit of a polymeric molecule.

monosaccharide   A single, or simple, sugar. A term generally referring to a subunit of a polysaccharide.

Morgan, T. H.   (1866–1945) A geneticist whose classic studies on the genetics of the fruit fly, *Drosophila melanogaster,* confirmed the Mendelian laws of transmission of traits from parent to offspring and gave rise to the concept of the gene. He was awarded the Nobel Prize in medicine in 1933.

morphogenesis   The developmental process by which a specific form and structure take place.

morphogens   Certain factors of maternal origin present in an egg and that help to determine the location, in the developing embryo, where limbs and other structures of the mature organism will appear.

morphology   The study of morphogenesis.

mosaicism   The property of a tissue being a mixture of clonal populations of cells. By the LYON EFFECT, only one X

chromosome of the pair will be expressed in any cell lineage; therefore, if cells with different X chromosomes activated are found in a cell population, it indicates that the cell population is a mixture of cell clones. Conversely, if cells are found in which the same X chromosome is always active, it indicates the cell population is of clonal origin. Such analysis has demonstrated that most tumors probably derive from a single cell.

***mos* oncogene**   An oncogene found in a RETROVIRUS that causes sarcoma tumors in mice. The name is derived from *Moloney Sarcoma virus.*

M phase   The period of the cell cycle covering mitosis. It is divided into prophase, prometaphase, metaphase, anaphase, and telophase.

M-phase promoting factor (MPF)   A protein factor, isolated from eggs of the frog, *Xenopus laevis,* that forces cells at any stage in the cell cycle into mitosis (M phase).

msr, msd   Loci that code for an unusual RNA–DNA hybrid molecule in which RNA transcribed from *msr* is covalently linked to *msd* DNA. This type of structure was first discovered in myxobacteria but has also been discovered in the soil bacterium *Stigmatella aurartiaca.*

muD phage   A variant of the BACTERIOPHAGE mu that has been engineered as a VECTOR to be used for determining PROMOTER activity using the β-galactosidease gene as a REPORTER GENE.

multidrug resistance genes (*mdr* genes)   Genes that confer resistance to the lethal effects of certain drugs, particularly chemotherapeutic agents, such as methotrexate. They arise through massive amplification of a single-copy gene and are present in homogenously staining regions or double-minute chromosomes. (See DOUBLE MINUTE.)

multinucleate   Having more than one nucleus within the same cytoplasm. (See HETEROKARYON, SYNCITIUM.)

mung bean nuclease   An enzyme that catalyzes the breakdown of single-stranded DNA into single nucleotides and short oligonucleotides that have phosphate groups on their 5'-ends.

murine   Of, or pertaining to the genus *mus,* which includes mice, rats, and other rodents.

murine leukemia virus (MuLV)   A retrovirus that causes leukemia in mice and carries the *abl* oncogene.

murine sarcoma virus   Also known as the Moloney sarcoma virus. The retrovirus carrying the *mos* oncogene.

mutagen   Any chemical, physical, or biological agent causing permanent, heritable alterations in the base sequence in DNA.

mutagenesis   The process of introducing mutations by exposing cells to a mutagen(s).

mutagenesis,   site-directed   A technique for introducing specific changes in base sequence in a cloned segment of DNA, usually by replacing a portion with a synthetic oligonucleotide.

mutagenesis *in vitro*   Mutagenesis of cells in culture that may differ from mutagenesis using the same mutagen in the intact organism in terms of the type and number of mutations.

mutations, somatic   Mutations that occur in the cells of the body as opposed to the germline, or reproductive, cells. Somatic mutations are not passed to an organism's offspring, whereas germline mutations are.

mutator loci   Genes whose function is associated with the fidelity of DNA synthesis; for example *mutD* = a subunit of DNA polymerase III, *mutU* = DNA

helicase, *mutY* = endonuclease, which cleaves mismatches between A and G.

mutator   phenotype   Phenotype with a strong tendency to undergo mutation. The mutator phenotype is expressed by any organism that carries a mutation in a gene involved in maintaining the fidelity of DNA synthesis.

myasthenia gravis   An autoimmune disease of the nervous system characterized by a progressive paralysis of the motor nerves. The disease is caused by the formation of antibodies to one's own neurotransmitter receptors acting at the neuromuscular junction.

mycelium   A mass of hyphae.

mycobacteria   A group of rod-shaped acidophilic bacteria including the bacteria that cause leprosy and tuberculosis.

mycology   The study of molds (fungi).

mycoplasma   The smallest independently growing organisms known. Mycoplasmas were isolated and characterized on the basis of their role in causing a type of pneumonia. Also known as pleuropneumonia-like organisms (PPLO).

mycotoxin   Any toxic substance produced by a fungus or mold. Among the mycotoxins used in molecular biological research are muscarine, phalloidin, ergotamine, and various antibiotics.

myelin   A class of lipids that comprises the outer sheath surrounding the axon of brain neurons. In this location it serves as an electrical insulator without which nerve impulses could not be propagated.

myeloblastosis (*myb*) oncogene   Oncogene carried by the avian myeloblastosis virus, causing myeloblastic leukemia in chickens.

myelocytomatosis (*myc*) oncogene   Oncogene carried by the avian leukosis retrovirus, causing a type of leukemia in birds.

Actin and Myosin Filaments in Muscle

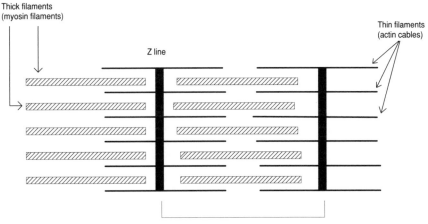

Thick filaments
(myosin filaments)

Thin filaments
(actin cables)

Z line

Sarcomere

myeloid cell   The collective term for all classes of blood cells not including T- and B-lymphocytes (i.e., all bone-marrow-derived blood cells).

myeloma   A tumor of the antibody-secreting lymphocyte cells. Also known as plasmacytoma.

myeloma proteins   An antibody molecule of the Ig class secreted by a myeloma tumor. Each myeloma antibody represents the specificity of a single antibody cell (monoclonal antibody).

myoblast   Embryonic cells that fuse with one another to form mature muscle cells.

myoglobin   The protein that carries and exchanges oxygen for CO_2 in muscle tissue. Like hemoglobin, myoglobin carries oxygen on a heme group at-

tached to polypeptide globins, but unlike hemoglobin is present only as a monomer.

myosin   One of the two contractile proteins of muscle. Myosin bundles interdigitate with actin bundles, and muscle contraction is the result of the two proteins sliding over one another.

myxobacteria   Bacteria that normally live in the soil as individual cells but that, under conditions where nutrients become limiting, form multicellular aggregates similar to primitive multicellular organisms.

myxovirus   A class of viruses identified and characterized on the basis of their role in causing influenza. Its two families are orthomyxoviruses and paramyxoviruses.

N

N-acetylneuraminic acid (NANA)
The molecule attached to a glycoprotein that forms the red blood cell membrane

receptor for influenza virus. This accounts for the ability of the virus to cause HEMAGGLUTINATION.   Neuraminidase cleaves off the NANA residue, thereby inactivating the receptor.

nalidixic acid   A synthetic antibiotic that mimics the action of the natural antibiotic novobiocin.

nanogram   0.000000001 or 10^{-9} gram.

nascent protein   A polypeptide in the process of being synthesized on RIBOSOMES.

native conformation   The three-dimensional shape that a biomolecule naturally assumes in its normal biological environment.

natural killer (NK) cells   A type of thymocyte (T-lymphocyte) found in the spleen and responsible for a particular immune response to various tumor cells.

negative complementation   The suppression of a normal gene by a mutant of that gene.

negative regulation (negative regulators)   A term applied to regulation of gene expression based on repression, rather than stimulation, of gene expression. In negative regulation, the target gene is normally expressed to its maximal extent and, under appropriate environmental conditions, expression is controlled by lowering its level of expression. (See LAC OPERON.)

negative selection   The selection of cells or organisms within a population expressing a desired trait by eliminating members of the population not expressing the trait. Negative selection is commonly used to select microorganisms expressing a gene, such as resistance to an antibiotic, that allows them to survive in the presence of that antibiotic while other microorganisms are eliminated that do not have, or express, the gene. (See HYPOXANTHINE–AMINOPTERIN–THYMINE SELECTION.)

neomycin   A synthetic antibiotic derived from streptomycin. Neomycin and other related antibiotics act on the RIBOSOME to inhibit bacterial protein synthesis.

neoplasia   Any abnormal growth of an adult tissue.

Nernst equation   An equation that relates the free-energy change for a reaction $R_1+R_2+R_3+\cdots+R_n \longrightarrow P_1+P_2+P_3+\cdots+P_n$, where R_i represents a reactant and P_i represents a product, to the concentrations of the reactants and products:

$$\Delta G=\Delta G^\circ+2.303RT \log \frac{([R_1][R_2] \cdots [R_n]}{[P_1][P_2] \cdots [P_n])},$$

where R = universal gas constant, T = temperature in kelvins, ΔG° = a constant for a given reaction.

nerve growth factor (NGF)   A factor, originally isolated from a tumor transplanted into a chicken embryo, that causes selective outgrowth of sensory and sympathetic neurons. NGF is essential for proper growth and survival of these types of neurons during development.

nested deletion   A deleted region of a nucleic acid occurring within a region covered by a second, larger, deletion of the same nucleic acid. The smaller deletion is said to nest within the larger deletion.

neu oncogene   An oncogene isolated from rat cells by transfection into 3T3 cells. Unlike the retroviral oncogenes, *neu* is not carried by a retrovirus. The *neu* proto-oncogene is activated to a cancer-causing (i.e., oncogenic) form by a point mutation. The nucleotide sequence of the *neu* oncogene is homologous to the *erb-B* oncogene.

neural cell adhesion molecule (N-CAM)   A protein present on the surface of neurons causing them to aggregate. N-CAM is expressed at specific times during development, suggesting that it plays a role in the development of neural structures, such as ganglia.

neuraminidase   A glycoprotein present as a spike on the outside of the influenza virus envelope. Neuraminidase acts to break down an inhibitor of the influenza virus hemagglutinin (HA) protein.

neurofilament proteins   Three proteins that are the subunits of neurofilaments, a type of intermediate filament in neurons. The function of neurofilaments is unknown.

neuron   The brain cell type that carries the nerve impulses involved in higher thought and movement.

neuropeptide   Any of a class of short polypeptides that function as neurotransmitters.

Neurospora   A mold used as a tool in genetic experiments because (a) it is normally HAPLOID, and (b) it forms a structure (the ascus) from which single-celled spores can be readily isolated. *Neurospora crassa* was the organism originally used to demonstrate that genes coded for individual proteins.

neurotransmitter   A chemical released from the axon of one neuron that induces a nerve impulse in an adjacent neuron when it binds to a specific receptor in the dendrite of the second neuron.

neutral substitution   A change in a nucleotide base in the coding region of a gene that does not produce a change in the activity of the protein.

neutrophil   One of three subclasses of white blood cells known as granulocytes or polymorphonuclear leukocytes (PMNs).   Neutrophils contain large multilobed nuclei and phagocytose small invading organisms such as bacteria. (See PHAGOCYTOSIS.)

nick   A gap in the sugar-phosphate backbone of a nucleic acid.

nick translation   A technique for labeling double-stranded DNA fragments to be used as HYBRIDIZATION PROBES. DNA polymerase is used in the presence of labeled deoxyribonucleotides to fill in nicks produced by treatment with low levels of DNase I.

Nick Translation

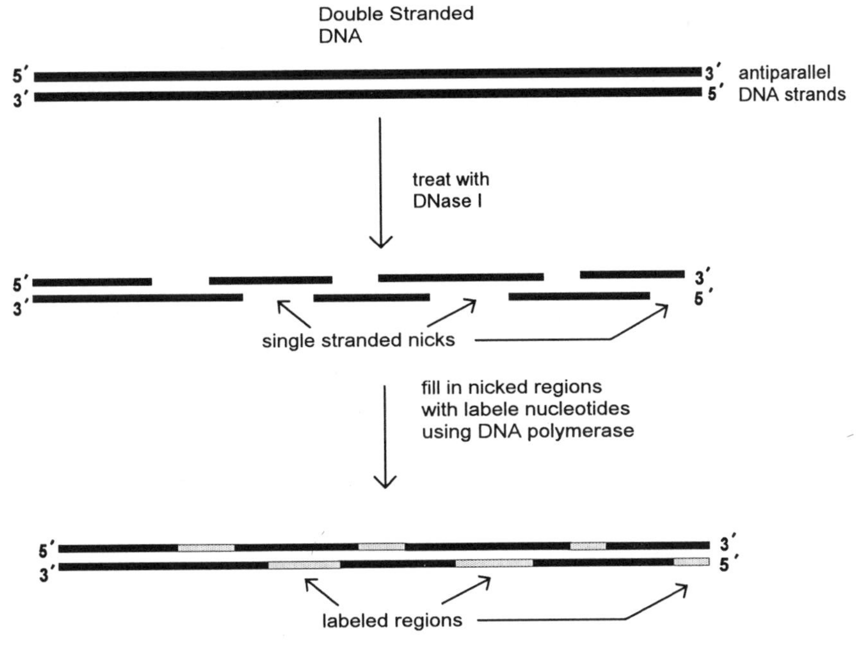

Double Stranded
DNA

5′ 3′ antiparallel
3′ 5′ DNA strands

treat with
DNase I

5′ 3′
3′ 5′

single stranded nicks

fill in nicked regions
with labele nucleotides
using DNA polymerase

5′ 3′
3′ 5′

labeled regions

Nicotamide Adenine Dinucleotide (NAD)

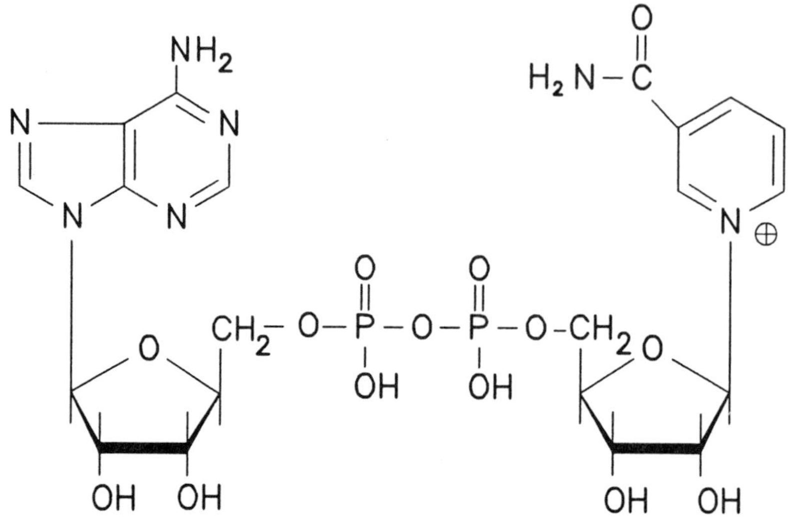

nicotinamide adenine dinucleotide phosphate (NADP) A cofactor for enzymes involved in oxidoreductions and electron transfer in numerous biochemical reactions, but particularly those involved in the oxidative metabolism of sugars for energy production. NAD is a combination of two nucleotides, one of which is derived from the B vitamin niacin.

nicotinic receptor One of two types of cholinergic receptors in brain cells; named because of its sensitivity to nicotine.

ninhydrin A chemical that forms a purple pigment after reacting with amino groups. This reaction is used in the detection and quantitation of proteins.

nitrifying bacteria Bacteria that carry out the process by which nitrogen gas in the atmosphere is converted to ammonia. This is the only means by which nitrogen can be incorporated into biologically useful molecules, such as proteins and nucleic acids. (See NITROGEN FIXATION.)

nitro blue tetrazolium (NBT) A chemical that forms a blue precipitate when certain substrates are acted upon by alkaline phosphatase. Hence, it is used as a colorimetric indicator in techniques utilizing ALKALINE PHOSPHATASE labels.

nitrocellulose filter A thin flexible membrane of nitrocellulose, a material that noncovalently binds tightly to a variety of biological materials, including proteins and nucleic acids. Because of this property, nitrocellulose is widely used as a binding material for carrying out COLONY HYBRIDIZATIONS and PLAQUE HYBRIDIZATIONS and Southern, northern, and western BLOTS.

nitrogen cycle The global process of nitrogen recycling. The nitrogen cycle involves the breakdown of organic materials with the liberation of ammonia and other inorganic nitrogen-containing compounds and the fixation of inorganic nitrogen in plants by the process of nitrogen fixation.

nitrogen fixation The process by which nitrifying bacteria convert nitrogen gas in the atmosphere into ammonia. Nitrogen fixation utilizes two systems: (1) the reductase, which generates electrons, and (2) the nitrogenase, which uses the electrons generated from the reductase to reduce nitrogen (as N_2) to ammonia (NH_3). (See RHIZOBIUM.)

nonautonomous controlling elements Defective transposons unable to transpose but that can do so when a normal TRANSPOSON is also present.

noncompetitive inhibition Inhibition of enzymatic activity by a substance that acts at a site on the enzyme different from the active site.

nondisjunction An error of chromosome segregation during cell division (either meiosis or mitosis) in which the daughter chromosomes (sister chromatids) fail to move to opposite poles.

nonessential amino acid An amino acid that can be synthesized from other amino acids or other precursors. For this reason the nonessential amino acids, in contrast to the essential amino acids, can be omitted from the diet without causing death.

nonidet P40 (NP40) A nonionic detergent (octylphenyl-ethylene oxide) that selectively breaks open the plasma membrane in animal cells but not the nuclear membrane. For this reason, NP40 is often used for rapid isolation of nuclei and cytoplasmic fractions of cells.

nonpolar group A small group of atoms held together by linkages where electrons are more or less equally distributed among the constituent atoms so that the linkages have no polar character. Such groups are important in biological systems because they are insoluble in water.

nonsense mutation A mutation that causes a normal codon to change into one of the three termination codons (TAA, TAG, or TGA). This type of mutation causes premature termination of synthesis of the protein in which it occurs.

nonheme iron Iron atoms found in iron-sulfur proteins as opposed to heme groups. The iron-sulfur proteins are electron carriers in the process of electron transport.

nonpermissive cell A host cell type that does not support the replication of a particular virus; for example, monkey cells are *permissive* for SV40 virus, but mouse cells are *nonpermissive* for SV40.

nonreciprocal recombinant chromosomes The result of RECOMBINATION between misaligned chromosomes such that a gene on one chromosome receives a duplication of part of the gene while the copy of that gene on the other chromosome suffers a deletion of the same material.

nonsense suppressor A mutant tRNA that permits the placement of an amino acid in a polypeptide when one of the nonsense codons is encountered during translation of an mRNA. Such tRNAs can reverse the effects of nonsense mutations (suppressor effects). (See MESSENGER RNA, TRANSFER RNA.)

nontranscribed spacer See SPACER DNA.

noradrenaline (norepinephrine) A chemical that is both a neurotransmitter and a hormone released by the adrenal glands. As a neurotransmitter, noradrenaline acts on certain neurons of the sympathetic nervous system. As a hormone, it acts on cells of the liver and muscle to stimulate sugar mobilization and storage.

northern blot An analytical technique in which RNA is run on an agarose

Noradrenaline (norepinephrine)

gel, blotted onto a membrane where it is hybridized to a specific probe. This technique is particularly useful as a means of detecting the expression of a particular mRNA. (See HYBRIDIZATION, BLOT.)

novobiocin An antibiotic, produced by *Streptomyces niveus,* that acts by preventing ATP binding to the enzyme DNA gyrase, thereby stopping DNA synthesis in the infectious bacteria.

NTG vector A mammalian expression vector that combines the enhancer–promoter activities of the long terminal repeats (LTRs) from the Moloney murine leukemia virus with the bacterial-derived neomycin resistance gene (*neo*^{r}) as a selectable marker. (See APH.)

***ntrB, ntrC* proteins** Regulatory proteins in the control of expression of genes involved in nitrogen metabolism under conditions of nitrogen deficiency in bacteria. When nitrogen-containing compounds in the bacterial medium drop to low levels, *nitrB* becomes activated; the activated form of *ntrB* is responsible for phosphorylation of *ntrC,* which stimulates the gene transcription. (See NITROGEN FIXATION.)

nuclear lamina A thin matrix of dense filaments just below the envelope surrounding the cell nucleus.

nuclear lamins The intermediate filaments that make up the nuclear lamina.

nuclear magnetic resonance (NMR) An analytical technique based on the absorption of radio waves by the atoms in solution, which depends on the magnetic properties of the individual atoms; the extent of absorption and the frequency at which absorption occurs (the NMR spectrum) provides information about the type and quantity of the atoms present. In biological preparations, the absorption characteristics of a molecule, (i.e., its NMR spectrum) are influenced by its three-dimensional structure and by its biochemical environment.

nuclear membrane (nuclear envelope) A double-walled membrane that forms the enclosure surrounding the nuclear compartment. The outer membrane is joined to, and may actually be thought part of, the endoplasmic reticulum.

nuclear pore complex An octagonal array of large protein granules that surround pores perforating the double-layered nuclear membrane at various points. The nuclear pore complex is a specialized channel through which nucleic acids and other materials shuttle between the nucleus and the cytoplasm.

nuclear scaffold A fibrous network that extends from the inside of the nuclear membrane throughout the nucleus and is attached to the cellular DNA at specific sites. The nuclear scaffold is seen by electron microscopy in isolated nuclei carefully treated with nucleases and either salt or detergents to remove histones, some nonhistone proteins, and the free (i.e., unattached) DNA strands from the CHROMATIN. The function of the nuclear scaffold is unknown but is believed to play a role similar to the chromatin itself in control of gene expression.

nuclear transplantation A technique for removing the nucleus from one cell and placing it into the foreign cytoplasm of a second cell (i.e., an enucleated cell).

nuclease Any of a class of enzymes that catalyze the breakdown of a nucleic acid(s).

nucleic acid A molecule of either DNA or RNA.

nucleohistone (histone) Any of the five different proteins that make up a nucleosome, designated H2A, H2B, H3, H4, and F1.

Nucleosomes

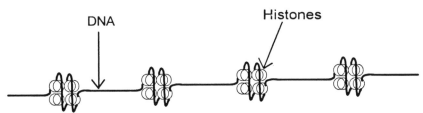

DNA

Histones

nucleoid body   The analog of the nucleus in bacteria. The nucleoid body, which contains the bacterial genomic DNA, is not enclosed in a membrane and the DNA is not complexed with chromatin, but is distinguishable as a large centrally located mass, appearing less dense than the surrounding cytoplasm.

nucleolus   A large structure within the nucleus of eukaryotic cells consisting of numerous loops of chromatin-bound DNA and containing clusters of tandemly repeated ribosomal RNA genes. The nucleolus is therefore the structure responsible for the production of rRNA and continually engaged in high levels of synthesis of rRNA.

nucleophilic group   Any cluster of covalently linked atoms that tends to donate electrons in a chemical reaction. Nucleophilic groups often initiate important biochemical reactions by attacking electron-deficient carbon atoms attached to oxygen atoms.

nucleoprotein   particles   Complexes of protein and RNA, primarily in the nucleus, that play a role in the processing of RNA.

nucleor organizer region   A cluster of rRNA genes on a DNA loop in the nucleolus.

nucleoside   A ribose or deoxyribose molecule attached to any purine or pyrimidine base via the first carbon atom of the sugar. Common nucleosides found in DNA and RNA are (deoxy) adenosine,

(deoxy) cytidine, (deoxy) guanosine, thymidine (deoxy form only), and uridine.

nucleoside antibiotic   An antibiotic that is a nucleoside, containing an analog of a purine or pyrimidine base. It acts by inhibiting the normal mechanisms of DNA and RNA synthesis in rapidly growing microorganisms. Nucleoside antibiotics (e.g., cytosine arabinoside), unlike other antibiotics, are active against viral infections.

nucleosome   An octameric structure complexed around a strand of DNA. Nucleosomes are evenly spaced along the DNA strand, forming a linker (nonhistone complexed) region of 142 base pairs. Nucleosomes are composed of two each of the different H HISTONES and are believed to play a role in regulating the expression of the genes they are complexed with.

nucleosome phasing   A model of nucleosome structure in which a certain DNA sequence is always located at a certain position on the nucleosome. If this can be shown to be true, it implies that some mechanism exists for aligning nucleosomes with certain sequences on the DNA in the nucleosome complex.

nucleotide   A nucleoside with a phosphate group attached either to the fifth or third carbon of the ribose or deoxyribose sugar.

nucleus   The central, membrane-enclosed structure containing the cell DNA in the form of chromatin in eukaryotic cells.

null DNA The DNA representing genes that are only expressed in single cell or tissue type. Null DNA is presumed to represent genes for specialized proteins unique to a specific cell type (e.g., hemoglobin genes in red blood cells).

null mutation A mutational event resulting in the complete elimination of a gene.

nurse cell Accessory cells in the ovary that surround an oocyte and supply a variety of macromolecules and nutrients to the oocyte via cytoplasmic bridges. Ribosomes, mRNAs, and proteins are passed to insect oocytes in this way.

nystatin An antibiotic with a polyene structure (i.e., containing many carbon-carbon double bonds) active against fungal infections. Nystatin is active against *Candida* infections when applied topically.

O

O-antigen A branched polysaccharide attached to a specific lipid (lipid A) on the outer surface of the cell envelope of the pathogenic bacterium *Salmonella typhimurium* and other gram-negative bacteria.

obligate anaerobe Bacteria that have an absolute requirement for oxygen and are not fermenting (e.g., tuberculosis).

ochre codon The nonsense codon, TAA, which is a signal for termination of polypeptide synthesis.

ochre mutation Any mutation that produces an ochre codon in place of a codon for an amino acid.

Okazaki fragment A short (1000–2000 bp) DNA fragment produced during DNA replication of the LAGGING (5′-terminating) STRAND of the template DNA. Okazaki fragments are initiated by a short RNA primer hybridized to the lagging strand and destroyed after synthesis of the Okazaki fragment is complete. (See HYBRIDIZATION.)

oligonucleotide A short strand of either DNA or RNA with a length of 2 to about 30 bases. The term generally refers to synthetic polynucleotides.

oligopeptide A polypeptide of anywhere between approximately 2 and 10 amino acids.

oligosaccharide A chain of sugars containing anywhere between approximately 2 and 10 monosaccharide subunits.

onc function The property acquired by a PROTO-ONCOGENE of inducing a cancer or promoting tumorigenesis when it becomes an oncogene.

oncogene The activated form of a proto-oncogene. Mechanisms by which proto-oncogenes become activated include transduction by a retrovirus, mutation, and chromosomal translocation, whereby the proto-oncogene is placed into a new genetic environment.

oncogenic Pertaining to any agent —chemical, physical, or biological—that causes cells to undergo changes characteristic of cancer cells.

oncogenic virus Any of a broad range of viruses that cause cells to undergo changes characteristic of cancer cells or to cause tumors in animals. (See ONCOGENE, RNA TUMOR VIRUS.)

ontogenetic   Of, or pertaining to, ontogeny, the complete life cycle or process of development of an organism.

oocyte   The DIPLOID germ cells of the female that generate gametes (eggs) by meiotic division. (See MEIOSIS.)

oogamy   The union of gametes to produce an embryogenic cell, as in fertilization of an egg by fusion with sperm.

oogenesis   The process by which the mature egg(s) is generated from an oocyte.

oogonium (oogonia)   The female reproductive organ in which the eggs are formed in thallophyte plants.

open reading frame   Any nucleic acid segment whose codons specify a continuous polypeptide; a nucleic acid segment without stop CODONS.

operator   A portion of the operon PROMOTER that acts as a regulator of operon expression by serving as a binding site of a repressor protein.

operon   A cluster of contiguous bacterial genes all under the control of a single promoter.

opiate   Any of the chemical derivatives of opium.

optical density   The property of absorption of light by a solution of any given substance. The decrease in intensity of a light beam at a certain wavelength, as it passes through a solution over a certain distance, is proportional to the molar concentration of the solution. (See BEER–LAMBERT LAW.)

optical isomer   See STEREOISOMER.

organelle   Any subcellular, membrane-enclosed structure in the cytoplasm of a eukaryotic cell that carries out a specific cellular function (e.g., mitochondria, chloroplasts, endoplasmic reticulum, Golgi, etc.).

ornithine   An intermediate in the urea cycle, the series of reactions in which nitrogen in the form of urea is formed. Ornithine is derived from the amino acid arginine through the loss of urea.

orotic acid   A pyrimidine base that is the common precursor of CTP and UTP, the triphosphate nucleotides of cytosine and uracil.

Orphan Drug Act   An act of Congress directed toward rare human diseases (defined as having a prevalence of less than 200,000 cases) that grants, as an incentive, a seven-year period of marketing exclusivity to the developer(s) of therapeutic drugs.

orphans   Genes that are members of a gene family but are in distant locations.

orthophosphate   Any salt of orthophosphoric acid (H_3PO_4); the name given to phosphoric acid stripped of one or more of its hydrogen atoms as the result of its being placed in aqueous solution. Orthophosphate is the form of phosphorous present in most important phosphorous-containing biomolecules (e.g., nucleic acids).

osmolality   The concentration difference between osmotic compartments as measured by molality; a 1 molal solution

Phosphoric Acid (phosphate)

$$HO - \overset{\overset{\textstyle O}{\|}}{\underset{\underset{\textstyle OH}{|}}{P}} - O$$

is defined as 1 mole of the solute dissolved in 1000 grams of the solvent.

osmosis   The spontaneous diffusion of a substance from a compartment of relatively high concentration to a compartment of relatively low concentration where, generally, the two compartments are separated from one another by a semipermeable membrane. Many nutrients and other substances of biochemical importance enter or exit cells by osmosis.

osmotic pressure   A pressure produced on the side of a membrane with a higher solute concentration caused by the passage of water across the membrane by osmosis.

osteoblast   The cells that initiate the formation of bone by secretion of the bone matrix, a material composed largely of collagen, which is hardened into bone by the deposition of calcium phosphate crystals.

oubain   A toxic glycoside that specifically inhibits the Na^+–K^+ ATPase. The use of this inhibitor provided important information that helped elucidate the functioning of the ionic pump.

oxidative phosphorylation   The formation of ATP from ADP (+phosphate) using the energy of the electron transport process to drive the reaction.

oxidizing agent   Any chemical agent that takes electrons, either as electrons or as electron-rich atoms, from another chemical with which it reacts.

oxidoreductase   The class of enzymes that carry out electron transfers between two substrates. Oxidoreductases are the enzymes responsible for many of the electron transfers that occur in the electron transport chain in which energy from the electrons derived from the metabolism of sugars is used for oxidative phosphorylation.

oxygenases   A group of enzymes that add oxygen across double bonds of the substrate molecule. This type of reaction is an essential step in the energy-producing metabolism of certain molecules (e.g., fatty acids).

P

p53   A human phosphoprotein of 53 kilodaltons. Originally discovered in studies of SV40 T antigen to which it is tightly bound, it was at first thought to represent the product of an oncogene. Recent evidence now indicates that p53 possesses antioncogenic, or tumor-suppressing, activity.

pair-rule mutants   Mutants of the fruit fly, *Drosophila melanogaster,* in which features of normal development are missing in every other SEGMENT. (See SEGMENTATION.)

palindrome   A nucleic acid base sequence that is a "mirror image" of itself; for example, the base sequence GTGGCCGGTG is a palindrome because it consists of the sequence GTGGC and its "reflection" CGGTG. The recognition sequences of most restriction endonuclease enzymes are palindromes.

papilloma virus   A member of the papova group of DNA viruses that produces generally benign tumors (papillomas) of the epithelial cell layer in rabbits, cattle, and humans. Recently discovered members of the subclass representing human papilloma viruses (HPV) are now believed to cause some malignant genital cancers.

papilloma, polyoma, and vacuolating viruses (papova) Describes a loose classification of viruses containing the papilloma, polyoma, and SV40 viruses. The classification is based primarily on a common morphology exhibited by these viruses.

paranemic joint A side-by-side nonhelical arrangement of DNA strands (as opposed to the usual double-helix relationship) that occurs when a single-stranded circular DNA undergoes *recA*-mediated recombination with a linear double-stranded DNA. (See *recA*.)

parasegments An alternative scheme for labeling the segments of the fruit fly, *Drosophila melanogaster;* each parasegment begins in the middle of a SEGMENT. This scheme gains its utility from the fact that some mutants of *Drosophila* are more easily visualized in terms of parasegments. (See SEGMENTATION.)

parasexual The recombination of genetic material from different individuals that differs from sexual reproduction in that the genetic material is not derived from specialized meiotic cell types (e.g., sperm and egg). Parasexual reproduction is characteristic of yeast.

parthenogenesis The process of reproducing without fertilization (i.e., asexually).

partition coefficient For some particular substance dissolved in two different solvents that do not mix with each other, the partition coefficient is the ratio of the amount of the substance that remains dissolved in one solvent to the amount of the substance that remains dissolved in the other solvent when the two solutions are mixed and then allowed to separate.

passive hemagglutination A test in which the presence (or absence) of an antibody is detected by the ability of the antibody to cause red blood cells to clump together (hemagglutination). Passive hemagglutination differs from the usual hemagglutination test because the surface of the red blood cells must be artificially modified by chemical linkage of a protein (the antigen) in order for the hemagglutination to take place.

passive immunity Immunity transferred from one individual to another by injection of blood components (e.g., blood serum or cells).

passive transport Movement of a substance across a membrane from one side where the substance is at a relatively high concentration to the other side where the concentration is relatively low. (See OSMOSIS.)

Pasteur, Louis (1822–1895) French chemist who demonstrated the principle of sterilization, thereby destroying the idea that life could arise spontaneously from nonliving organic material (spontaneous generation).

Pasteur effect The observation that when a microorganism living under anaerobic (oxygen-free) conditions is suddenly exposed to oxygen, sugar consumption as well as accumulation of the sugar-breakdown product, lactate, drops.

pasteurization The process of destroying disease-causing microorganisms by heat. (See STERILIZATION.)

patent A document certifying an inventor or inventors to exclusive rights to an invention and any profits that may be derived from its use, sale, or license. A number of the products of biotechnology, such as transgenic mice and various therapeutic products derived from recombinant DNAs, have been awarded patents. However, the application of patent law to some areas of biotechnology, such as the patenting of individual genes, remains controversial.

pathogen Any microorganism causing disease or producing a pathological condition.

pathway A series of biochemical reactions occurring in a specified sequence by which a particular molecule (the pre-

Penicillin Backbone

penicillin derivatives
obtained by substituting
groups for "R"

$$R - \underset{\underset{O}{\|}}{C} - HN - C - C \overset{S}{\diagdown} C \overset{CH_3}{\underset{CH_3}{\diagup}}$$

$$O = C - N \longrightarrow C - COOH$$

cursor) is modified to become another, usually for purposes of synthesizing essential biochemicals or for degrading them.

Pauling, Linus (b. 1901)   A chemist who studied the behavior of electrons in chemical bonds. He won the Nobel prize in chemistry in 1954 for his studies on the nature of electrons in bonds between the amino acids in proteins. This discovery helped elucidate the structure of the α-helix form in proteins.

pBR322   A commonly used PLASMID for cloning recombinant DNAs in the bacterium *E. coli.*

pectin   A jellylike substance released by plants. Pectin consists of chains of the sugar derivative galacturonic acid.

pectinase   An enzyme that degrades pectin by breaking the links between its sugar units.

P-element(s)   A type of transposable element found in the fruit fly, *Drosophila.* (See TRANSPOSON.)

pellet   The sediment portion of a biological extract after the extract is subjected to centrifugal force.

penicillinase   An enzyme that inactivates penicillin by breaking a key bond in the penicillin molecule through hydrolysis.

penicillins   Products of the *Penicillium* molds that act as an antibiotic by destroying bacteria. They interfere with the cross-linking of proteins in the bacterial cell wall, causing the bacterium to break open, or lyse. The penicillins are all derived from a common chemical backbone, the lactam ring, by substituting various chemical groups given the general designation 'R' at a certain position on the ring. See LACTAM ANTIBIOTICS and 6-AMINOPENICILLIC ACID.

pentose   Any sugar with a five-carbon-atom backbone.

peptidases   A class of enzymes that break down proteins by cleaving the peptide bonds between the individual amino acids making up the protein (e.g., the digestive enzymes chymotrypsin, trypsin, and pepsin). The breaking of peptide bonds by peptidases occurs by hydrolysis.

peptide   A group of amino acids covalently linked by peptide bonds in a linear chain.

peptide antibiotic   A short peptide with antimicrobial properties (e.g., gramicidin A).

peptide bond   A covalent linkage between the $-NH_2$ group of one amino acid and the $-COOH$ group of another amino acid. This type of linkage is known as an amide bond when applied to links between molecules that are not amino acids.

peptide hormone   A short peptide secreted into the bloodstream that in-

A Heptapeptide

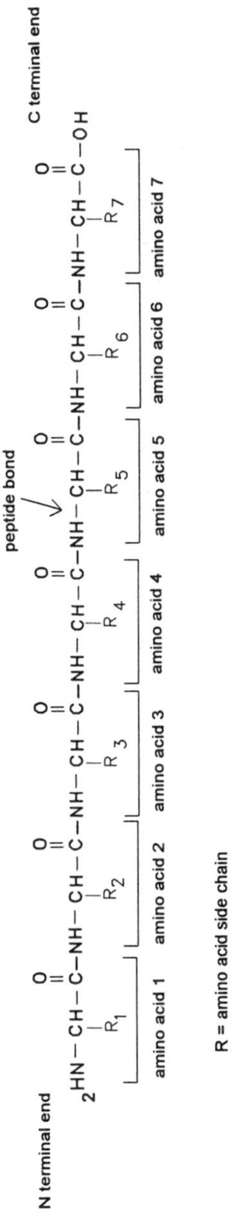

N terminal end

peptide bond →

C terminal end

$$_2HN-CH-C-NH-CH-C-NH-CH-C-NH-CH-C-NH-CH-C-NH-CH-C-NH-CH-C-OH$$

with each C bearing $=O$, and each CH bearing side chains R_1, R_2, R_3, R_4, R_5, R_6, R_7

amino acid 1 | amino acid 2 | amino acid 3 | amino acid 4 | amino acid 5 | amino acid 6 | amino acid 7

R = amino acid side chain

---

duces biological activity in a distant target gland or organ. Examples are the pituitary hormones, such as growth hormone (GH), thyroid stimulating hormone (TSH), and adrenocorticotropic hormone (ACTH).

peptidoglycan   A peptide covalently attached to chains of sugars or sugar derivatives. Peptidoglycans are structural components of bacterial cell walls. In gram-positive bacteria the peptidoglycan portion of the cell wall is present in many layers.

peptone   The water-soluble portion of a protein that has been partially broken down (i.e., hydrolyzed), such as by boiling. (See HYDROLYSATE.)

perfusion culture   A culture in which there is a continual inflow of fluid carrying nutrients or other substances.

pericentriolar material   Material of unknown composition surrounding the centriole of the chromosome. This material serves as the anchor points for the microtubules, which pull the chromosomes apart during MITOSIS.

perinuclear space   The space between the inner and outer nuclear membranes.

periodicity   In a structure with a regularly repeating subunit, periodicity refers to the distance that represents one complete subunit.

permissive host   A cell that, infected by a virus, allows the expression of a particular viral function(s), usually replication of the virus.

peroxidase   An enzyme that acts to promote the breakdown of hydrogen peroxide (H_2O_2) into water according to the reaction

$$H_2O_2 + \text{substrate-}H_2 \xrightarrow{\text{peroxidase}} 2H_2O + \text{substrate}_{ox}.$$

peroxidase labeling   The attachment of a peroxidase enzyme (e.g., horseradish peroxidase) to a probe so that the presence of the probe can be visualized by a colorimetric reaction based on the enzyme activity.

peroxisome   Small, self-replicating cytoplasmic organelles, containing no DNA, composed largely of the peroxidase enzyme catalase.

petite mutant   A microorganism (especially yeast and euglena) lacking mitochondria. In yeast such mutants form tiny colonies when grown on a nutrient source low in sugar. (See MITOCHONDRION.)

P-factors   DNA sequences carried on various chromosomes in the male fruit fly that bring about hybrid dysgenesis in matings with females of certain strains (called M strains for maternal contributing).

pH   A measure of the acidity of an aqueous solution; if the pH value of a solution is below 7, the solution is acid, whereas solutions with a pH value above 7 are alkaline.

phage   Short form of bacteriophage (literally, ''bacteria eater''). A virus that infects a bacterium and then usually destroys it. The destruction of the infected bacterium (the host) with the release of progeny viral particles is the last step in the virus life cycle.

phagocyte   Any cell that normally carries out phagocytosis; usually applied to certain white blood cells, for example, macrophages that carry out phagocytosis as part of their function in the immune system.

phagocytic index   An assay for the detection of phagocytic activity in a blood specimen. Among the tests used for this purpose are staining by the dye nitro blue tetrazolium (NBT), which turns blue when particles containing the dye are phagocytized, or the uptake of latex beads by phagocytic cells.

phagocytosis   The process by which a particle or cell becomes engulfed and ultimately devoured by another cell for sustenance or defense.

phagosome   Following phagocytosis, the engulfed particle is found in a membrane-enclosed vesicle, referred to as a phagosome, in the cytoplasm of the phagocyte.

phalloidin   An alkaloid derived from the toadstool *Amanita phalloides* that binds to the actin filaments in a cell, thereby preventing cell movement.

pharmacology   The study of the action of drugs particularly as it relates to their therapeutic uses.

phase-contrast   microscopy   A type of microscopy in which the image of the specimen being viewed is enhanced by a technique involving manipulation of the light deflected by the specimen. In normal microscopy only the light passing straight through the specimen is used to create the image of the specimen.

phenotype   The set of characteristics that make a living organism distinct from others.

phenylalanine   One of the 20 amino acids that make up proteins; designated as Phe or F. Phenylalanine is also used in the body to make the neurotransmitter dopamine as well as adrenaline.

phenylketonuria (PKU)   A genetic disease based on an inability to convert the amino acid phenylalanine into the amino acid tyrosine. This results in the accumulation of a toxic substance (phenylpyruvate) that causes severe mental retardation if the condition, which is manifest in newborns, is not treated by adherence to a diet low in phenylalanine. Since the disease has been traced to a deficiency of a particular enzyme (phenylalanine hydroxylase), prevention of the disease in susceptible individuals is a goal of modern genetic engineering.

pheromone   A chemical signal, secreted by an animal, that brings about a specific behavior (e.g., mating) in an animal of the same species.

Philadelphia chromosome A type of reciprocal translocation between chromosomes 9 and 22 seen in patients with chronic myelogenous leukemia (CML).

phloem A type of plant vascular tissue surrounding the xylem that makes up the vessels conducting fluids downward along the stem or trunk of the plant toward the root.

phosphatide A type of phospholipid made up mainly of glycerol, fatty acids, and phosphate. Phosphatides are the type of lipid making up the bulk of the phospholipids found in cell membranes.

phosphatidylethanolamine (PE) A type of phospholipid common to many cell membranes. PE is also a common constituent of many artificial membranes, such as those used to construct LIPOSOMES.

phosphodiesterase An enzyme that catalyzes the breakage of phosphodiester bonds.

phosphodiester bond A covalent bond that attaches a phosphate group to any other group by an oxygen atom bridge, such as sugar$-$O$-$PO$_3$$=$. Phosphodiester bonds are the linkages that join the sugar molecules to one another in the backbone of nucleic acids.

phospholipase Any of a class of enzymes that acts to break down phosphatides by breaking the bonds between the glycerol portion of the phosphatide and the attached fatty acid(s) or the bonds between the glycerol portion and the phosphate.

phospholipid The general class of lipids, made up of fatty acids and phosphate, that are the main component of all cell membranes. Phosphatides and sphingosines are the two major phospholipid subclasses.

phosphomycin A phosphate-containing antibiotic produced by *Streptomyces*.

phosphoramidite A chemically modified nucleotide used in the synthesis

A Phosphoramidite

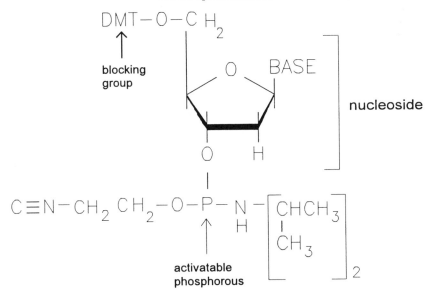

of oligonucleotides containing an activated phosphoester group at the 3'-carbon and a DIMETHOXYTRITYL (DMT) blocking group at the 5'-carbon.

phosphoribosyltransferase An enzyme necessary to make nucleotides from free purine and pyrimidine bases. The action of this enzyme is responsible for various important biomedical and research applications, such as labeling of nucleic acids or killing cancer cells with lethal analogs of the normal purines and pyrimidines.

phosphoric acid See ORTHOPHOSPHATE.

phosphorylation Any of a variety of biochemical processes by which a phosphate group is added to an organic molecule. However, the term usually applies to the phosphorylation of nucleosides, particularly adenosine, resulting in the formation of adenosine mono-, di-, and triphosphates (AMP, ADP, and ATP respectively). Phosphorylation resulting in ATP formation is the major means of storing energy for all forms of biological activity. (1) *Photophosphorylation* ATP formation produced during photosynthesis and therefore light requiring. (2) *Oxidative phosphorylation* Oxygen-dependent formation of ATP. This means of generating ATP is coupled to the process of sugar oxidation (i.e., breakdown) in animal cells. (3) *Substrate level* Generation of ATP occurring as part of the oxidation of sugars but without requiring oxygen.

photoaffinity labeling The use of light to activate certain light-sensitive molecules so that they spontaneously bond to a protein, nucleic acid, or other type of molecule. This technique is used for labeling biologically important substances if the activated molecule provides a highly visible color or generates some other type of signal with high detectability.

photoautotroph A photosynthetic organism capable of living on only minimal nutrients—in other words, capable of making all its necessary biomolecules from simple organic molecules.

photoheterotroph A photosynthetic organism deficient in the ability to make one or more of its essential biomolecules. It therefore requires nutritional supplements in its growth medium.

photomultiplier A photosensitive device that uses photoemission and secondary electron emission to detect low levels of light. Electrons emitted from a photosensitive material by incident light are accelerated and focused onto a secondary-emission surface (called a dynode). Several electrons are emitted from the dynode for each primary electron produced. The secondary electrons are then directed onto a second dynode where more electrons are released. This process is repeated several times to amplify the initial photocurrent so that extremely low levels of light can be detected.

photon The unit of light that represents one discrete packet of light energy.

photoreactivating enzyme See DNA PHOTOLYASE.

photosynthesis The process by which plants utilize light energy to create sugars and produce oxygen from carbon dioxide and water.

photosystem A cluster of chlorophylls and other pigments that functions to capture the light energy used to carry out photosynthesis.

phototroph An organism dependent on light for nourishment via photosynthesis.

phragmoplast The enlarged football-shaped spindle seen toward the end of MITOSIS in plant cells; the structure in which the cell plate forms.

phycomycetes A class of primitive fungi with many features in common with fungi.

phytochrome A pigment protein believed to play a role in the initiation of plant development when activated by light in the red or near-red part of the spectrum.

phytohemagglutinin A class of proteins causing clumping of red blood cells (hemagglutination) by binding to certain sugar chains on the cell surface; also referred to as lectins. Examples are concanavalin A and ricin.

phytotoxin Any of a number of highly poisonous substances produced by plants.

picogram 10^{-12} or 0.00000000001 gram.

Picornaviridae (picornaviruses) A class of RNA viruses originally termed enteroviruses because they were discovered in the intestinal tract. Picornaviruses are classified as enteroviruses (poliovirus, coxsackievirus, echovirus, and enterovirus) and rhinoviruses (rhinovirus). The name is derived from *pico* (small) and *rna* to denote RNA.

pinocytosis A variation of PHAGOCYTOSIS in which the engulfed particle is taken into the cell in small vacuoles representing pinched-off pieces of the original particle.

piperidine A chemical that causes breakage of the sugar–phosphate backbone in nucleic acids at any point along a nucleic acid strand at which the purine or pyrimidine rings have been partially oxidized or completely removed. Piperidine treatment is used to create subfragments of a larger nucleic acid in the Maxam–Gilbert procedure for sequencing nucleic acids.

pituitary gland A small endocrine gland located at the base of the brain

that secretes important polypeptide hormones, including follicle stimulating hormone (FSH), leutinizing hormone (LH), and prolactin, all of which play a role in stimulation of the female reproductive organs. The pituitary gland also produces adrenocorticotropic hormone (ACTH), somatotropin, and thyrotropin.

pK, pK$_a$ Terms representing the "strength" of a chemical reaction; the degree to which some reaction will proceed in the direction written, for example, as A+B→C+D. The negative logarithm of the equilibrium constant ($-\log[K]$, where $K=[C][D]/[A][B]$) for the chemical reaction.

plankton The small floating plant and animal life in a body of water.

plaque A clear area in an immobilized carpet of bacteria produced by local destruction of the bacteria in that area by BACTERIOPHAGE.

plaque assay A means of determining the number of bacteriophage in a suspension by counting the number of plaques produced in a certain amount of the suspension. The results are usually expressed as plaque-forming units per milliliter of suspension (PFU/mL).

plaque hybridization A process in which a labeled probe is annealed to the DNA from bacteriophage in a plaque. Plaque hybridization is used to identify plaques containing bacteriophage-carrying recombinant DNA.

plasma The liquid portion of blood.

plasma cell An antibody-secreting white blood cell.

plasma gel The protoplasm of a protozoan in the gel form, such as the protoplasm in the pseudopodium of *Amoeba proteus*.

plasma membrane The membrane surrounding the cytoplasm of a eukaryotic cell.

plasma sol   The protoplasm of a protozoan that is not in gel form.

plasmid   A piece of DNA in the cytosol of bacteria that replicates independently from the bacterial chromosome. Naturally occurring plasmids have been found to carry a number of genes, perhaps most importantly the genes that confer resistance to a number of antibiotics. Genetically engineered plasmids are important vectors for carrying recombinant DNAs.

plasminogen activator   An enzyme deriving its name from the ability to catalyze the conversion of a plasminogen to plasmin that then catalyzes the breakdown of fibrin, a major component of blood clots. The secretion of plasminogen activator is a marker of cell transformation to a cancerous or precancerous state. Plasminogen activator is used as a therapeutic agent to dissolve blood clots associated with blockage of the coronary arteries.

plastid   An organelle in plant cells containing its own genome; chloroplasts are a type of plastid.

platelet   A subcellular particle in blood that is actually a fragment of a megakaryocyte cell formed in the bone marrow. Platelets bound to fibrinogen initiate the formation of a blood clot at the site of a wound.

platelet-derived growth factor (PDGF)   A growth factor in the granules of platelets that probably plays a role in wound healing. PDGF consists of two subunits, one of which was found to represent a slightly changed version of the *sis* oncogene.

β-pleated sheet   Rigid, extended sheetlike secondary structure of proteins held together by hydrogen bonds.

plectonemic   Pertaining to the standard double-helical arrangement of double-stranded DNA. (See DOUBLE HELIX.)

pleiotropic   Any agent, such as a hormone, having more than one effect or having an effect on more than one target.

pleomorphic   Having variable form, such as variations in shape, behavior, or other characteristics of organisms of the same species.

point mutation   A change in a single nucleotide in a gene, resulting in loss of function or altered functioning of that gene.

POL   One of the three major genes of retroviruses. The POL gene encodes the protein for the viral enzyme reverse transcriptase.

polar body   A small cell that is produced as a result of uneven separation of cytoplasm during MEIOIS when an oocyte is produced; the larger of the daughter cells becomes the oocyte.

polar group   A small group of atoms held together by dipole-dipole linkages.

polarimeter   An instrument for determining the percentage of polarized light in a beam of light and whether polarized light is rotated after passage through crystals of a given compound.

polarity   (1) Applied to a molecular bond between two atoms, refers to a state in which the electrons in the bond are localized more to one atom than the other, giving that atom a partial negative charge (while the other atom is partially positively charged). The presence of polar bonds confers important chemical properties to the compound containing them, including solubility in water. (2) Applied to cell microanatomy, polarity refers to specialization of the cell architecture at different parts of the cell, for example, the presence of cilia and secretory vesicles located at the apical, as opposed to the basal, end of the epithelial cells lining the gut and respiratory tracts. (3) Applied to nucleic acid strands, the term refers to the fact that the two ends of any nucleic acid strand are distinguishable from one another by whether the end is 5' or 3'. This gives the strand a directionality or polarity.

polar microtubules The microtubules extending between the polar bodies that have the apparent function of pushing the poles of a dividing cell apart during MITOSIS.

polar mutation A nonsense mutation that causes early termination of normal transcription in a gene and which therefore also prevents transcription of any subsequent genes in a POLYCISTRONIC unit.

poliovirus A picornavirus that infects individuals via ingestion but then attacks the central nervous system, resulting in varying degrees of paralysis.

polyacrylamide A polymer of acrylamide,

$$CH_2{=}CH{-}\overset{\displaystyle NH_2}{\underset{\displaystyle |}{C}}{=}O.$$

A gel is formed when the polyacrylamide strands are cross-linked. Polyacrylamide gels are used for electrophoresis of proteins and nucleic acids.

polyacrylamide gel electrophoresis Separation of nucleic acids or proteins from a heterogeneous mixture on the basis of size or charge by placing the mixture in a polyacrylamide gel and subjecting it to an electric field.

polyadenylated The general term for nucleic acids that have undergone polyadenylation.

polyadenylation The addition of long tracts of adenosine polymers to the tail ends (the 3'-ends) of messenger RNAs in eukaryotic cells.

adenosine   adenosine
   | |
····-phosphate-ribose-phosphate-ribose-

adenosine
   |
phosphate-ribose-····

polyadenylation signals Certain sequences of bases in an RNA molecule required for polyadenylation to occur; for example, the sequence AAUAAA located in a region 11–30 nucleotides from the end of an mRNA molecule.

polyamine Molecules formed from repeating hydrocarbon chains separated by amino groups. The chain is always terminated at each end by a positively charged amino group. Because of their positive charge, the polyamines function to stabilize nucleic acids by neutralizing the strong negative charge of the nucleic acid phosphate backbone. Putrescine, spermidine, and spermine are the common polyamines.

poly-A polymerase The enzyme responsible for polyadenylation of an RNA strand.

polycistronic A region of a nucleic acid containing sequences representing multiple genes (cistrons) in an end-to-end tandem arrangement.

polycistronic mRNA The messenger RNA transcribed from a polycistronic DNA.

polyclonal antibody A set of antibodies, secreted by a corresponding set of antibody-producing white blood cells. Although each of the antibodies carries a unique specificity, the set of antibodies as a whole reacts with a variety of antigenic molecules.

polyelectrolyte A large molecule highly charged under biological conditions.

polyethylene glycol (PEG)   A long polymer of

$$-\overset{\displaystyle OH}{\underset{\displaystyle |}{C}}-$$

groups. PEG is used in inducing cell fusion, precipitating microscopic particles, dehydrating samples of biological materials, and stabilizing certain enzymes.

polylinker   A synthetic polynucleotide containing the sequences representing the restriction sites of certain specified restriction enzymes. (See RESTRICTION ENDONUCLEASE.)

polymer   A chain comprising many identical or different molecular units (monomers); a network of similar linked molecules.

polymerase chain reaction (PCR)   A technique for rapid amplification of extremely small amounts of DNA using the heat-stable Taq I DNA polymerase enzyme. PCR has found wide application in forensic medicine because analyzable quantities of nucleic acid can be obtained even from microscopic tissue samples.

polymerases   A class of enzymes that catalyzes the formation of long nucleotide polymers, particularly as a means of making template-driven copies of nucleic acids (See DNA POLYMERASES, RNA POLYMERASES.)

polymorphism   A naturally occurring variation in the normal nucleotide sequence within the individuals in a population.

polymyxin   A group of antibiotics derived from the bacterium *Bacillus polymyxa* with activity primarily against gram-negative bacteria.

polynucleotide   A polymer consisting of a long chain of nucleotides. The 3'-end carries an unreacted hydroxyl group (—OH group) on the 3'-carbon; the 5'-end carries an unreacted phosphate group on the 5'-carbon.

polynucleotide kinase   An enzyme that catalyzes the transfer of a phosphate group from ATP to the free 3'-end of a polynucleotide.

polynucleotide ligase   An enzyme that links two polynucleotides together. The enzyme catalyzes the formation of a covalent bond between a phosphate group on the 5'-end of one polynucleotide and the free 3'-hydroxyl group at the end of the other polynucleotide.

polyoma   A member of the Papova group of viruses, which normally infects rodent cells. (See PAPOVA VIRUS.)

polypeptide   A polymer of amino acids; the term usually applies to a peptide chain of less than 100 amino acids.

polyploid   Having more than the normal diploid number of chromosomes.

polyploidy   The state of being polyploid.

polyribosome (polysome)   A number of RIBOSOMES attached to the same mRNA.

polysaccharide   A chain of sugar molecules linked end to end.

polyspermy   Fertilization of a single egg by more than one sperm.

polytene chromosomes   Giant chromosomes found in insect salivary gland cells. They are actually made up of thousands of copies of the DNA normally found in one chromosome.

polyteny   The state of a cell containing polytene chromosomes.

poly uridylic acid (poly U)   A polymer of uridine of undefined length.

P/O ratio   The amount of phosphate incorporated into ATP divided by the amount of O_2 taken up by the MITOCHODRION during RESPIRATION. Generally taken as a measure of the efficiency of

The Polymerase Chain Reaction

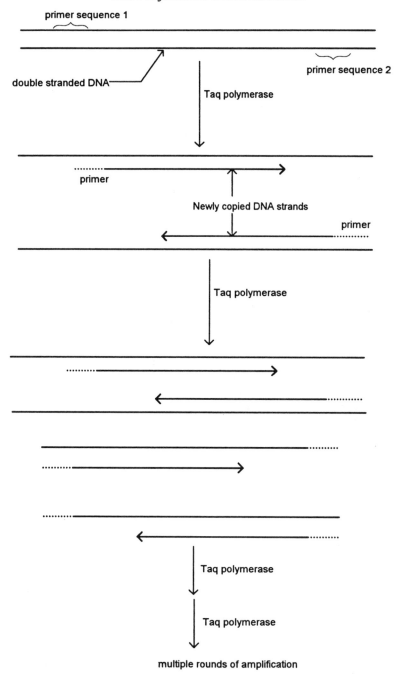

energy production when sugars are oxidized.

porphyrin An organic molecule made up of four nitrogen-containing rings called pyrroles. Modified porphyrins are the basic constituents of the active sites of hemoglobin, myoglobin, chlorophylls, and cytochromes.

position effect The influence of the gene position on its activity, such as is seen when a gene moved (e.g., by translocation) to a new chromosomal location becomes inactive.

post-transcriptional processing Certain specific changes in the RNA that occur before the RNA leaves the cell nucleus in mature form. Polyadenylation, capping, and splicing are examples of post-transcriptional processing.

post-translational import A process by which a certain class of proteins is brought into the interior of the endoplasmic reticulum (the ER) either following or during its synthesis on RIBOSOMES that are bound to the ER; polypeptides that are to be imported are recognized on the basis of the fact that they contain a small sequence of amino acids known as a signal peptide at one end.

post-translational modification Some alteration in the structure of a polypeptide, such as addition of a polysaccharide chain (glycosylation), after it is synthesized and usually after it is imported into the interior of the ENDOPLASMIC RETICULUM. Such modification is required for the polypeptide to take on its biological activity.

post-translational processing Removal of a specific end piece of a polypeptide known as the signal peptide, following its synthesis on ribosomes; one part of the process of post-translational import.

post-translational transfer The import of a polypeptide into the membrane of the endoplasmic reticulum or an organelle after synthesis of the polypeptide (i.e., translation) is completed.

poxvirus A class of DNA viruses that produce transient inflammatory skin lesions (e.g., chickenpox, smallpox, etc.).

precursor A substance from which another substance is made by a series of sequential changes in molecular structure.

prednisone A synthetic steroid hormone used to reduce chronic inflammation such as occurs in arthritis.

pre-mRNA The general term given to that subclass of RNAs present in the nucleus that will be processed to become mature messenger RNA (mRNA) but that have not yet undergone that processing. (See POST-TRANSCRIPTIONAL PROCESSING.)

prenatal diagnosis The diagnosis that a disease exists in a developing fetus made on the basis of examination of cell or tissue samples taken from the fetus in the womb.

preproinsulin A precursor of insulin. Proinsulin, the immediate precursor of insulin, is produced from preproinsulin by cleavage resulting in the removal of a short polypeptide portion.

preprotein The polypeptide precursor of a membrane-bound protein prior to its actual insertion into a membrane. Since the leader sequences of membrane-bound proteins are removed upon insertion into the membrane, preproteins have leader sequences.

Pribnow box A sequence of bases in the DNA making up part of the gene *promoter*. The Pribnow box always occurs at 10 base pairs from the site at which transcription starts and consists of the sequence TATAAT or a close variation.

primary culture Cell culture arising from a tissue specimen when it is first placed into culture.

primary response The elicitation of an immune response to a foreign antigen after an animal is first exposed to the antigen.

primary structure The sequence of amino acids making up a polypeptide.

primase The enzyme that catalyzes the formation of short RNA primers required to copy the DNA strand starting from the 5'-end.

primeosome A complex of proteins including DNAG PRIMASE wound around a hairpin fold of single-stranded DNA. This complex is the structure in which synthesis of the RNA primers in Okazaki fragments takes place. (See BACTERIO-PHAGE $\phi \times 174$.)

primer A short oligonucleotide that ANNEALS to a specific region on a DNA or RNA strand and is used by a polymerase as a place to begin synthesis of a complementary nucleotide strand.

primer extension A technique of mapping genes in which a primer is annealed to a DNA or RNA fragment and then extended using an RNA or DNA polymerase (e.g., the KLENOW FRAGMENT of DNA polymerase I) and the four nucleoside triphosphates to copy the nucleic acid to which the primer is annealed. Primer extension is most commonly used to detect mRNAs containing the primer sequence.

priming The process of annealing a primer.

probe Any oligonucleotide containing a chemical label allows the oligonucleotide to be traced when it is annealed by hybridization to some target nucleic acid. To a lesser extent, any biomolecule, including a protein, lipid, or polysaccharide, that binds to some target molecule and bears a chemical label that can be traced after binding has occurred.

processed   pseudogenes Pseudogenes that show a close similarity in nucleotide sequence to the mRNA for their active counterparts. The existence of processed pseudogenes has been taken as evidence that some pseudogenes were somehow originally derived from mRNAs.

profilin A protein that complexes with actin proteins, preventing the polymerization of these proteins into actin filaments.

progesterone A steroid hormone produced by the ovary that prepares the uterus for reception of the fertilized egg.

prokaryote A term for the family of all primitive organisms (e.g., bacteria) in which the cellular DNA is not enclosed in a nucleus.

prolactin A hormone that stimulates the production of milk by the mammary glands, following pregnancy.

prometaphase The stage preceding metaphase in which chromosomes newly formed in the cytoplasm begin to migrate to the center of the cell where homologous chromosomes will pair with one another during metaphase. (See MITOSIS.)

promoter A sequence of bases in a nucleic acid strand that serves as a signal for the start of transcription of a given gene.

prophage The genome of a BACTERIOPHAGE that has become integrated into the chromosome of the host bacterium.

prophase The first phase of MITOSIS characterized by the appearance of chromosomes from the amorphous chromatin.

prostaglandins   A class of hormonelike chemicals derived from fatty acids. The more than 16 prostaglandins are classified into nine groups. Stimulation of muscle contraction and inflammation are believed to be caused by prostaglandins. The anti-inflammatory effects of aspirin are related to inhibition of prostaglandin synthesis.

prosthetic group   An organic molecule, often containing metal atoms, tightly bound to an enzyme and required for the enzyme function.

protease   The class of enzymes that catalyzes the cleavage of a polypeptide or protein into smaller polypeptides.

protein   Any polypeptide or cluster of polypeptides with a defined biological function.

protein hydrolyzate   The partial breakdown product produced by heating a protein mixture or subjecting it to treatment with acid or proteases.

protein kinase   The class of enzymes that catalyzes the transfer of a phosphate group from one compound (e.g., ATP) onto a protein.

protein kinase C   A protein kinase embedded in the cell membrane and believed to be activated by tumor promoters. Activated protein kinase C is then believed to cause transformation to a cancerous state by phosphorylation of as yet unidentified protein(s).

protein synthesis   The process by which amino acids are assembled into peptides on RIBOSOMES using the information supplied by a messenger RNA. Proteins and nucleic acids are held together by bonds susceptible to hydrolysis, and their assembly is accomplished by a reversal of the hydrolytic reaction. (See TRANSLATION.)

proteoglycans   Complexes between GLYCOSAMINOGLYCANS and protein that function as support matrices for skin and connective tissue, such as cartilage.

proteolysis   Breakdown of proteins by cleaving the bonds between amino acids by the process of hydrolysis. (See PROTEASE.)

prothrombin   A blood protein acted upon to form thrombin.

prothymocytes   Immature T cells formed from stem cells in the bone marrow prior to the time they enter the thymus. Prothymocytes are recognizable by the presence of an incomplete set of T-cell-receptor proteins on the cell surface.

proton gradient   An uneven distribution of protons caused by the accumulation of protons on one side of a membrane. Proton gradients are a means of storing energy for synthesis of ATP in mitochondria and chloroplasts. (See ADENOSINE TRIPHOSPHATE.)

proton-motive force   The amount of energy stored in a proton gradient.

proton pump   A cluster of membrane-embedded proteins that transports protons from one side of a membrane to the other.

proto-oncogene   A normal cellular gene that, when altered in a particular fashion (activation), acts to induce a cancerous state.

protoplast   A plant cell or bacterial cell in which the cell wall has been removed (e.g., by treatment with lysozyme).

protoplast fusion   A technique for introducing foreign or genetically engineered DNA into a cell by fusion of that cell with a protoplast carrying the DNA of interest.

prototroph   An organism with the same nutritional requirements as the parent organism from which it was derived.

protozoa   A phylum of single-celled organisms, such as *Paramecium caudatum* or *Amoeba proteus,* representing the most primitive animals (from *proto=* first and *zoan=*animal).

provirus   The name given to DNA representing the genome of a virus that has become integrated into the DNA of the host it infects.

pseudogene   A version of a gene that has become inactive over time as a result of accumulated mutations.

pseudouridine   An unusual form of uridine found only in TRANSFER RNAS (tRNA).

psoralens   A type of organic molecule that spontaneously forms a variety of covalent bonds with nucleic acids in the presence of ultraviolet light.

psychrophile   An organism thriving at low temperatures.

psychrotroph   An organism requiring low temperature for normal growth.

Ptashne, Mark (b. 1940)   Discoverer of the function cI repressor protein, which controls the state of lysogeny in the lambda bacteriophage.

pulsed-field gel electrophoresis   A variation of agarose gel electrophoresis allowing the separation of extremely large (several thousand kilobases in length) DNA fragments by agarose gel electrophoresis.

pulvomycin   An antibiotic that acts by blocking the elongation of a polypeptide chain as it is being synthesized. Pulvomycin interacts with an elongation factor (EFTu) and prevents the formation of an essential complex between GTP and an aminoacyl tRNA.

purine   A nitrogen-containing, double-ringed organic molecule that is the

Purine Nucleotide

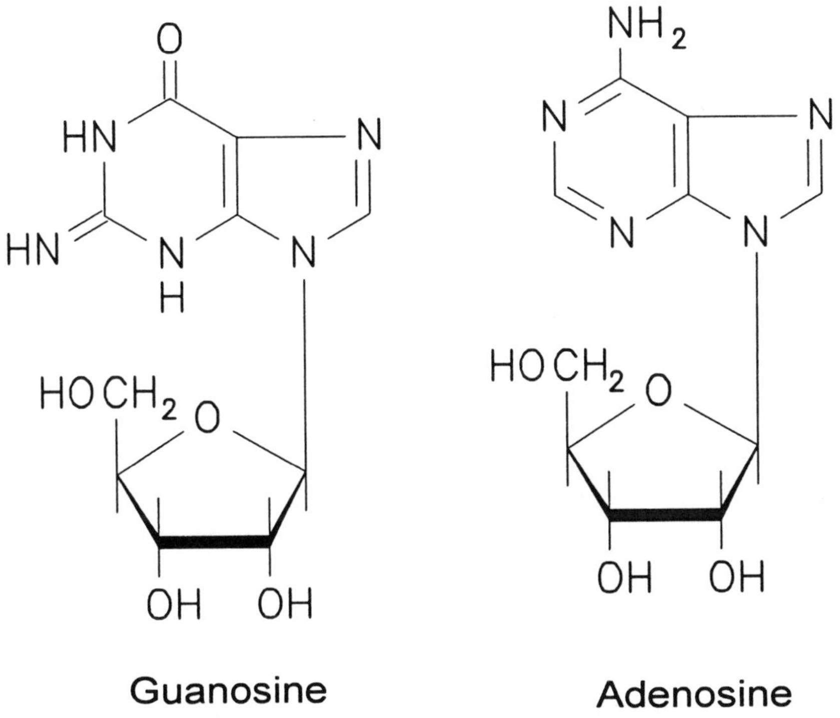

Guanosine

Adenosine

Purine Ribonucleosides

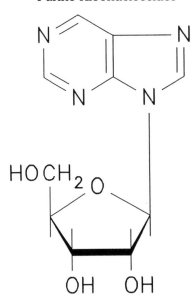

parent compound for the purines found in nucleic acids. (See APPENDIX III.)

purine bases in nucleic acids The purine-derived molecules adenine and guanine. In nucleic acids these molecules are attached to the sugar ribose or deoxyribose in the nucleic acid backbone.

purple bacteria A type of bacterium capable of carrying out photosynthesis.

pyridine An organic molecule containing five carbon atoms and one nitrogen atom in a ring. Used to dissolve otherwise difficult-to-solubilize biological materials.

pyrimidine A six-membered, nitrogen-containing, ringed molecule that is the parent compound for the pyrimidines found in nucleic acids. (See APPENDIX III.)

Pyrimidine Ribonucleosides

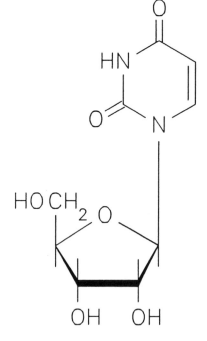

Cytidine   Uridine

pyrimidine bases, in nucleic acids
The pyrimidine-derived molecules thymine, uracil, and cytosine. Like the purine bases in nucleic acids, these molecules are attached to the sugar ribose or deoxyribose in the nucleic acid backbone.

pyrogen   Any of a number of toxic, fever-causing agents, usually of bacterial origin, such as endotoxins. The presence of pyrogens is a main concern when preparing solutions for injection since sterilization may destroy live bacteria but not their residual pyrogens.

Q

Qa locus   One of the genetic subloci within the mouse major histocompatibility locus (H-2). Qa codes for an antigen only found on a subset of blood cells, and so it is considered to be a differentiation antigen.

q banding   The technique of staining chromosomes with quinacrine, producing a unique pattern of chromosomal bands that can be used clinically for chromosome identification.

queuosine   An unusual purine base found only in transfer RNA. Queuosine is formed by adding a pentenyl ring to 7-methylguanosine.

quinacrine   A synthetic antimalarial compound also used as a fluorescent stain for chromosomes.

quinine   An alkaloid drug derived from the bark of the cinchona tree found in South America and Indonesia. Quinine is an antimalarial drug also used to relieve fever and pain in other diseases. At one time quinine was the only drug available for treatment of malaria, but it has been replaced to a large extent by synthetic drugs, such as quinacrine.

quinone   A class of cyclic organic compounds widely used to carry hydrogen atoms in certain critical steps in the process of energy production in both plants and animals; examples are phylloquinone, plastoquinone, and ubiquinone. Chemically, quinones are characterized by the presence of two keto groups ($C=O$) on the same hydrocarbon ring.

R

racemate   A mixture of two different forms of a molecule that do not differ from each other chemically but have a different physical arrangement of atoms that can be distinguished by methods using polarized light.

radial immunodiffusion   An immunological test based on the reaction of an antibody with a protein that has been allowed to seep out of a central well into a slab of agar where the reaction takes place.

radiation, α-   A high-energy electromagnetic radiation produced during the process of nuclear decay in which the α helium is emitted.

radioimmunoassay   A sensitive test for a particular protein based on the reaction of that protein with an antibody specific for it and where one of the reacting agents is radioactively labeled.

raf oncogene   An oncogene in murine sarcoma virus associated with fibrosarcoma tumors in rodents and humans. The name derives from *rat fibrosarcoma*.

random primer labeling   A technique for labeling DNA based on the annealing (see ANNEAL) of a mixture of short primers with randomly determined sequences to the DNA strand. Labeling takes place by extension of the primers (see PRIMER EXTENSION) using labeled nucleotides.

***ras* oncogene**   The oncogene in *rat* sarcoma virus and associated with sarcomas in rodents and carcinomas in humans. In mammals the *ras* protooncogene has a GTP binding activity believed to be homologous to the action of G proteins. (See GUANOSINE TRIPHOSPHATE BINDING PROTEINS.)

rb   An abbreviation for the antioncogene protein product of molecular weight 110,000 coded for by the gene associated with the familial form of RETINOBLASTOMA.

R banding   A characteristic pattern of bands produced when chromosomes are stained by various dyes (e.g., olivomycin). The basis of the pattern of R bands is the abundance of DNA rich in quanine–cytosine (G-C) base pairs.

reading frame   Any of three possible ways of reading a sequence of amino acids from a nucleic acid using the genetic code; the three different reading frames are determined by which base in any group of three consecutive bases is chosen as the start point.

reassociation kinetics   A technique for estimating the number of copies of a particular nucleic acid base sequence in a sample by measuring the rate at which denatured nucleic acid in the unknown sample ANNEALS to strands of a known nucleic acid with the same or similar sequence to that being determined.

reassociation of DNA   Reannealing of the complimentary strands of DNA

Nucleic Acid Reading Frames

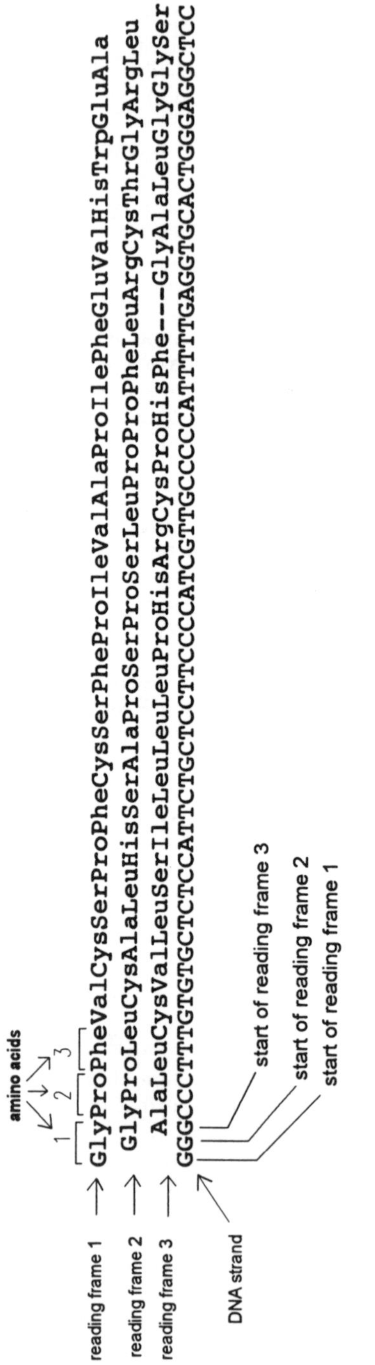

amino acids

reading frame 1 → GlyProPheValCysSerProPheCysSerPheProIleValAlaProIlePheGluValHisTrpGluAla

reading frame 2 → GlyProLeuCysAlaLeuHisSerAlaProSerLeuProSerLeuProProPheLeuArgCysThrGlyArgLeu

reading frame 3 → AlaLeuCysValLeuSerIleLeuLeuLeuProHisArgCysProHisPhe--GlyAlaLeuGlyGlySer

DNA strand → GGGCCCTTTGTGTGCTCCATTCTGCTCCTTCCCATCGTTGCCCCATTTTGAGGTGCACTGGGAGGCTCC

start of reading frame 3
start of reading frame 2
start of reading frame 1

after the paired strands have been separated by heat or strong acid or alkali.

recA A protein from the bacterium *E. coli* that causes the exchange of single strands of DNA between different double-stranded DNA molecules for recombination or repair of a defective tract of DNA.

receptor A specialized cell-surface molecule or complex of molecules that serves as a site of attachment for a specific effector molecule (the ligand), such as a hormone. The receptor may also function to produce a biological response in the cell to which the ligand is bound via its receptor.

recessive A form of a gene whose effect on the PHENOTYPE of an organism is masked by the alternative (dominant) form of the gene.

recessive allele The term for a recessive gene on a chromosome.

reciprocal translocation An exchange of material between chromosomes usually by breakage of each of the participating chromosomes at a specific site. A translocation involves movement of a physical portion of a whole chromosome.

recombinant DNA A DNA that has become joined to another, unrelated, or foreign, segment of DNA.

recombinase An enzyme that catalyzes the joining of immunoglobulin gene segments during the recombination event involved in immunoglobulin gene switching.

recombination The   process   by which DNA from a gene on a large-

Recombination

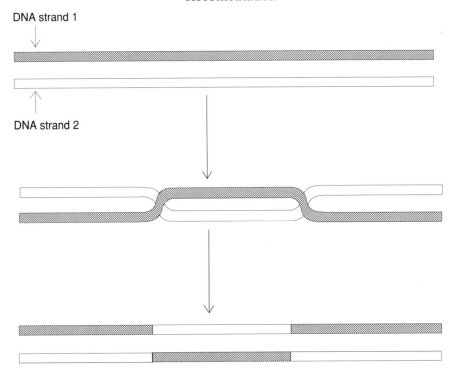

DNA strand 1

DNA strand 2

recombined DNA strands

genetic unit (e.g., a chromosome) becomes exchanged with the corresponding DNA on a complementary genetic unit, such as another chromosome that is an allele.

recombinational repair   A mechanism for repairing thymine dimers based on recombining an undamaged piece of DNA from the undamaged strand into the damaged region during DNA replication.

reconstituted   viral   envelopes (RVEs)   Viral envelopes whose contents have been removed or replaced with other substances. RVEs are made by a two-step process in which the whole virus is first completely disassembled, and then the components of the envelope portion are allowed to reassemble. RVEs are used to deliver substances of biological interest to various types of animal cells. (See DELIVERY SYSTEM, FUSOGENIC VESICLE.)

redox potential   A measure of the affinity that an atom or a molecule has for electrons. The redox potential is usually denoted by E_o', a number representing the extent to which the atom or molecule in question donates or accepts electrons to or from hydrogen as a standard reference.

redundancy   Existing in more than one copy, such as in repetitive DNA.

refractory phase   A time period required after emission of a nerve impulse for regeneration of a neuron's ability to emit a new nerve impulse.

regulatory enzyme   An enzyme that is part of a biochemical pathway, usually the first enzyme in the series, and serves as a regulator of the chemical reactions in the pathway by speeding up or slowing down the chemical reaction it controls in response to some environmental condition(s).

regulatory gene   A gene whose product controls the expression of another gene or genes. The repressor proteins of the lac operon and the bacteriophage cI regions are examples of regulatory gene products. (See LAC REPRESSOR PROTEIN.)

regulatory sequence   A nucleic acid sequence that serves as a site at which the protein product of a regulatory gene attaches; attachment of a regulatory protein to a regulatory sequence is the mechanism by which a regulatory gene controls the expression of another gene. (See ENHANCERS, SILENCERS.)

regulon   A set of spatially separated genes under the control of a single repressor-operator system. (See OPERATOR.)

relaxed   The state of a large molecule (e.g., a long DNA molecule or a protein) being in a loose conformation or loosely folded over on itself. Changes in biological activity are often related to the degree of twisting, folding, or compression of a molecule.

relaxin   A hormone released by the corpus luteum of the ovaries, causing relaxation of the pelvis to facilitate birth.

release factor   A protein that causes the process of translation to terminate when a termination signal on the mRNA is present.

rel oncogene   An oncogene in *reticuloendotheliosis* virus and associated with lymphatic leukemia in birds.

renaturation   The process of a molecule assuming its native shape or conformation after disruption of the native state (e.g., the reassociation of DNA).

reovirus   A group of RNA-containing viruses that infect the gut and respiratory tracts, usually without causing observable disease. The name derives from respiratory enteric orphan.

repair synthesis Synthesis of new DNA to replace a defective segment that has been removed; repair synthesis usually involves creating a complementary DNA strand from the remaining, nondefective DNA strand. (See EXCISION REPAIR.)

repetitive DNA A class of eukaryotic DNA sequences present in many, sometimes thousands or millions, of copies throughout the genome. (See ALU ELEMENTS.)

replacement sites DNA nucleotide bases that, when changed (e.g., by mutation), result in a change(s) in the amino acid(s) the DNA codes for.

replica plating A procedure by which bacterial colonies growing on one bacterial plate are reproduced on a second bacterial plate in the exact relative positions to one another as they were in the original plate.

replication fork The portion of the partially replicated DNA consisting of the separated DNA strands plus the

DNA Replication Fork

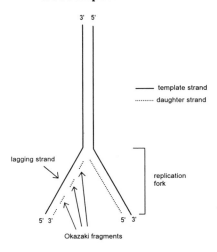

newly synthesized copies. See LAGGING STRAND, OKAZAKI FRAGMENT and ANTIPARALLEL.

replication of DNA The process in which DNA is copied by using each of the strands as a template for a new strand. Because with each round of DNA synthesis the new DNA consists of one newly synthesized strand and one parental strand, the process of DNA replication is called semiconservative. (See DNA POLYMERASE(S).)

replication origin A sequence of nucleotide bases that provides a signal for the start of DNA replication.

replicon The exact site on the DNA at which replication is actively taking place. (See REPLICATION ORIGIN.)

replicon fusion The meeting of two replicons approaching each other from opposite ends of replicating DNA.

reporter gene A gene whose expression is linked to the expression of another gene or biochemical process.

repressible enzymes An enzyme whose expression is governed by repression of gene transcription.

repressor protein A regulatory protein that exerts control over gene expression by binding to a regulatory sequence and thereby preventing transcription of the gene.

reptation The process by which a nucleic acid strand is threaded through the pores in an agarose gel matrix in a "headfirst" fashion during pulsed-field gel electrophoresis.

resolvase An enzyme that catalyzes the site-specific recombination event resulting in the integration of some transposable elements. Resolvase is coded for by a gene on a transposable element. (See TRANSPOSON.)

resorufin-β-D-galactopyranoside (RG)
A synthetic substrate for the enzyme β-galactosidase, the product of the LACZ gene, which produces a fluorescent product in the presence of the enzyme. RG is used to detect the expression of the *lacZ* gene in individual cells by fluorescence-activated cell sorting (FACS).

respiration   The oxygen-dependent process of generating energy in the form of ATP from sugars. (See ADENOSINE TRIPHOSPHATE, ELECTRON TRANSPORT.)

resting potential   The electrical potential across the membrane of a neuron between nerve impulses.

restriction endonuclease (restriction enzyme)   An enzyme, produced by bacteria, that cleaves DNA at a place defined by a specific sequence of nucleotide bases.

restriction fragment length polymorphism (RFLP)   The term is generally used to describe any pattern of restriction fragments that deviates from normal either in terms of the sizes or number of restriction fragments. RFLPs, which are observed in Southern blot hybridizations, are often used to study and analyze genetic diseases.

restriction mapping   A technique of mapping DNA by determining the location of sites for different restriction enzymes. By convention, each nucleotide base in the DNA sequence is numbered consecutively beginning at the 5' end, which is always the upper left end of the sequence when written in the form indicated by the asterik. This computer generated map names the restriction enzymes that have recognition sites along the sequence and indicates the base number at which the enzyme cuts (e.g., ACC 1,270). The two scales below the map show relative distances along the sequence in terms of base number and percentage of the total length. Shown below the map is a DNA sequence containing 650 nucleotides mapped by computer. Each letter in the sequence represents one of the purine or pyrimidine bases (A=adenine, C=cytosine, G=guanine, T=thymine). (See figure.)

restriction site   The nucleotide base sequence on DNA that specifies the site where a given restriction endonuclease will cleave.

reticuloendothelial system (RES)   All the phagocytic cells of the body except the circulating leukocytes. Cells of the RES remove particulate matter, such as foreign antibody-agglutinated antigens from the bloodstream. (See PHAGOCYTOSIS.)

retinoblastoma   A cancer of the retina cells occurring in small children. Retinoblastoma was one of the first cancers shown to run in families (familial retinoblastoma) and was therefore genetic in origin (See RB.)

retrovirus   A class of viruses characterized by having an RNA genome and carrying the enzyme reverse transcriptase within the virus capsid. The name was originally an acronym derived from *reverse transcriptase=retravirus* and later became retrovirus.

retrovirus vector   A genetically engineered DNA for cloning recombinant DNA that utilizes certain control elements from retroviruses.

reverse genetics   Type of genetic analysis in which the structure of a gene is determined from the protein it codes for. In the more common type of analysis the protein structure is determined from the structure of its corresponding gene.

reverse mutation   A mutation that reverses the effects of a previous mutation. The reverse mutation may or may not be localized near or at the site of the mutation whose effects it suppresses. (See SUPPRESSOR MUTATION.)

A Computer-generated Restriction Map

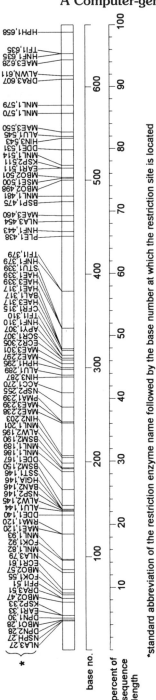

base no.

percent of
sequence
length

*standard abbreviation of the restriction enzyme name followed by the base number at which the restriction site is located

input sequence:

CACCATAGTTCTAATTTTCCCACATGCGATCAGGAGAGTAGTCCACCAAGTGGAAATAGAATTCTTCATCCTCCATGT
CCTCTATATCCATCTCTCTTTTCCATCCCACTCCACCTAGTTGGCTCTTTCTTGTCTGAGCTCTTGCCGAGACGGCTCT
TCCTGAGTTCCCTCCTCCAGTCTCTCCTCCTCTACTCCGTTGACTGCCAGATTGTCTTACAGCATAGATGAAACCACGTGAC
TTCTGTGCCCCAAGACTTTGATGTCTACAGAATAAAGTTCAAGCTTCTCAACGTTGTCACCTGGAATCTGGCCACAATTG
ATCTTTTCAGGCCTATCTCCCCTATCTCCTTTCTAATTACACATTTTGATTCTTGCCCAACCCAACCACTCACTATT
TCCAAGCACACCCTATACTTTCCCACGCCTTTGACTCCCACATGACCTTTGTCACACCTGCCTCCCTTCTGCTCTGCCTG
ACAAATTTTAACCTCTTCTTCAAACACCAGCCCAAATGCTCAGTTCCATAAAGCTTCTGTGACCTTGCCTCGCCTGCCTC
AGAGAGAAGTAATTTGCTTTTAGAGTTCACACAGTGCCTGTGAATACTTGTTGAGTGACTGAATCAACTTGCTCATAGCA
ATTTCATATT

reverse transcriptase   The enzyme, made and used by retroviruses during their life cycle, that catalyzes the synthesis of DNA copied from an RNA template. The enzyme is widely used in genetic engineering and molecular biology to make so-called complementary DNA (cDNA) from various RNAs so that base sequences in RNA can be cloned and manipulated by recombinant-DNA technology.

reverse transcription   The term that describes the action of the enzyme reverse transcriptase.

Rhesus blood groups   Classification of blood cells according to whether they react with antibodies to the blood cells of Rhesus monkeys.

rheumatoid factors   Certain antibodies (IgM), present in the blood of some individuals with rheumatoid arthritis, that react against other antibodies. Since it was discovered that different rheumatoid factors are specific for only certain subgroups of antibodies, rheumatoid factors became a means of classifying an individual's antibodies into subclasses (allotypes).

Rh factor   An antigenic substance on the surface of the blood cells of any individual that carries the Rh trait (Rh$^+$). The presence of the Rh factor in a fetus whose mother is Rh$^-$ may provoke a life-threatening agglutination of the fetal blood cells. The term is named for the Rhesus monkey, the organism used to demonstrate the presence of the antigen. There are at least 30 distinct subtypes of Rh factor.

rhinovirus   A picornavirus that infects the nasal cavity and causes many of the symptoms of the common cold, particularly nasal symptomatology.

Rhizobium   A leguminous plant that is a rich source of nitrogen-fixing bacteria living in large nodules attached to the plant roots.

rho factor   A small bacterial protein responsible for causing transcription to terminate when a RIBOSOME encounters an appropriate termination signal on the mRNA.

riboflavin   Vitamin B$_2$; an important cofactor for enzymes involved in the metabolism of sugars for ATP production. Riboflavin acts to transport electrons derived from the oxidation of sugars in energy production. (See ADENOSINE TRIPHOSPHATE.)

ribonuclease   A class of enzymes that breaks down RNA by breaking the bonds between phosphate and ribose molecules in the RNA backbone.

ribonucleic acid   A polymer of ribonucleotides, the purine and pyrimidine base sequence which, generally, is complementary to a DNA base sequence. There are four major classes of RNA performing different functions in the process of protein synthesis: messenger RNA (mRNA), transfer RNA (tRNA), ribosomal RNA (rRNA), and small nuclear RNA (snRNA).

ribonucleotide   A molecule consisting of ribose bound to phosphate and with a purine or pyrimidine base attached to the ribose molecule. Ribonucleotides are the building blocks of RNA.

ribose   A five-carbon sugar, used in ribonucleotides, normally in a ring conformation. (See CARBOHYDRATE.)

Ribose

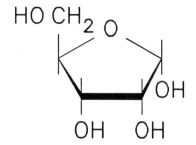

ribosomal protein Any one of many proteins that, together with a strand of RNA, form a ribosomal subunit. (See RIBOSOME.)

ribosomal RNA (rRNA) A long strand of RNA that together with ribosomal proteins, is a component of a ribosomal subunit. In eukaryotic cells there are two rRNAs denoted as 18s (found in the small ribosome subunit) and 28s (found in the large subunit). rRNA is thought to play a role in binding mRNA in the process of translation.

ribosome A small organelle in the cytoplasm that is the site where protein synthesis takes place. Ribosomes are made up of two subunits. The subunits are assembled with a strand of RNA to begin protein synthesis. (See RIBOSOMAL RNA.)

rickettsia Small bacteria that, unlike most other bacteria, are obligate parasites and live within, instead of outside, the cells of the infected host. Rickettsia are the agents that cause typhus fever and Rocky Mountain spotted fever.

rifampicin An antibiotic that blocks transcription by inhibiting the action of RNA polymerase in bacteria, specifically by inhibiting the initiation of the process of transcription.

R loops R loops are the segments of DNA representing INTRONS seen as single-stranded loops in electron micrographs of heteroduplexes between the eukaryotic mRNA and the genomic DNA from which the mRNA was transcribed.

RNA maturase An enzyme involved in transcript splicing in the yeast mitochondrial CYTOCHROME b gene. The RNA maturase gene is unusual in that part of the gene is found in an INTRON of the cytochrome b gene itself

RNA polymerases The class of enzymes that catalyze the synthesis of a strand of RNA, using DNA as a template to guide the assembly of ribonucleotides so that the order of the purine and pyrimidine bases in the DNA template is precisely copied, in complementary fashion, in the newly synthesized RNA.

RNA secondary structure The hairpin folding of an RNA molecule caused by internal base pairing of complementary stretches of purine and pyrimidine bases.

RNase D An exonuclease that removes nucleotides from the 3'-end of an RNA in one-at-a-time fashion. RNase D is involved in the maturation of tRNAs synthesized in large precursor strands shortened into functional tRNAs. (See TRANSFER RNA.)

RNA tumor virus A subclass of retroviruses that produces cancers by activation of oncogenes.

RNA–DNA hybrid(s) A double-stranded hybrid molecule in which RNA is base-paired with a complementary strand of DNA.

ros oncogene An oncogene found in a stain of avian sarcoma virus and associated with sarcoma tumors in birds. The name derives from Rochester 2 sarcoma virus.

rotavirus A class of RNA-containing viruses that infect the intestinal tract and are responsible for epidemic gastroenteritis and infantile diarrhea.

rough ER ENDOPLASMIC RETICULUM covered with attached RIBOSOMES. Proteins synthesized by the ribosomes on the rough ER are destined to be transported out of the cell via vesicles derived from the endoplasmic reticulum.

Rous sarcoma virus (RSV) A retrovirus that produces sarcoma tumors in chickens. RSV, discovered and named for Peyton Rous, was the first RNA tumor virus discovered.

R1 particle   An intermediate stage in the formation of one of the ribosomal subunits (the 30S subunit). An R1 particle is a complex formed from a strand of RNA and 15 ribosomal proteins.

R_0t   In the annealing (see ANNEAL) of RNA to DNA, a variable equal to the molar concentration of the RNA multiplied by time allowed for RNA–DNA annealing. R_ot values are generally used in plots of the annealing of RNA to complementary DNA sequences. (See C_0t VALUE.)

rubella   An RNA-containing virus (togavirus) responsible for German measles.

rumen bacteria   Bacteria living in the rumen of ruminant animals such as cows and utilizing urea that would otherwise be excreted to make amino acids, which are then returned to the animal's circulatory system.

Runting syndrome   A pathological condition, characterized by skin lesions, diarrhea, and death, that results when the lymphocytes from a mature animal are placed in, and then attack, the tissues of a newborn.

S

saccharide   Biochemical term for a sugar.

Saccaromyces cerevisiae   A yeast widely used as a vehicle for cloning extremely large segments of foreign DNA and for molecular studies on many animal genes that have homologues in yeast.

saline   A solution of sodium chloride at a concentration exactly equivalent to that found in bodily fluids (8 grams per liter; 0.8%).

Salmonella   A group of gram-negative, rod-shaped bacteria responsible for typhoid fever and a number of wide-ranging intestinal disorders.

saltatory movement   The directed movements of organelles in the cell cytoplasm. This type of movement is thought to be controlled by MICROTUBULES.

saltatory replication   Replication of a DNA sequence that produces extra copies of the sequence along the same

DNA strand. This type of process is believed to have been responsible for the highly repeated, tandemly arrayed sequences in satellite DNA.

salting-out   The phenomenon of causing dissolved proteins or nucleic acids to precipitate out of solution by the addition of salts.

salt stabilization   A phenomenon whereby slow denaturation of proteins and nucleic acids in aqueous solution is prevented by the addition of salts.

Sanger, Frederick (b. 1918)   Discoverer of the first means by which the amino acid sequence of a polypeptide could be determined. Sanger is famous for the discovery of the amino acid sequence of insulin in 1954; he was awarded the Nobel Prize in chemistry in 1956.

Sanger (dideoxy) sequencing   A technique for determining the sequence of a segment of DNA. The technique utilizes synthetic nucleotides (DIDEOXY-NUCLEOTIDES) to create small polynucleotides representing small subfragments of the DNA to be sequenced and that can be made to terminate specifically at any of the four purine or pyrimidine bases.

Sanger method   A method for determining a polypeptide sequence based on determination of the identity of the terminal amino acids of small subfragments of the original polypeptide.

saprotroph   An organism that obtains nourishment from nonliving matter.

sarcoma-derived growth factor (SDGF)   A growth factor secreted by cells infected with murine sarcoma virus RNA tumor virus. Since noninfected cells treated with SGF undergo changes generally characteristic of cells transformed into a cancerous state, sarcoma-derived growth factor is now referred to as transforming growth factor (TGF).

sarcoplasmic reticulum   A membranous structure that surrounds the myofibrils in muscle tissue. The sarcoplasmic reticulum contains calcium pumps that regulate the level of calcium ion (Ca^{2+}) in muscle tissue.

sarcosine   A component of the antibiotic actinomycin D, an inhibitor of transcription. Chemically, sarcosine is *N*-methyl glycine.

satellite DNA   A type of DNA made up mostly of repeated sequences not transcribed into RNA and found near the chromosome CENTROMERE.

satellite RNAs   See VIRUSOIDS.

scanning electron microscopy (SEM)   A variation of electron microscopy in which the specimen is given a thin coat of metal so that the electron beam can be used to visualize details of the cell surface as opposed to internal structures.

Scatchard analysis (plot)   A mathematical method for estimating both the number of receptors for a certain ligand and the affinity of the ligand for its receptor from a plot of the amount of unbound ligand versus the ratio of bound ligand to free ligand.

Schiff's reagent   A chemical (fuchsin leucosulfonate) used in the periodic acid-Schiff stain (PAS) to identify the presence of certain infecting microorganisms, such as fungi.

schistosomiasis   A group of diseases whose symtpoms range from dermatitis to cirrhosis of the liver. The symptoms are caused by parasitic infection by one of the trematode worms of the genus *Schistosoma*. Schistosomiasis is endemic in the populations of Africa, the Middle East, and South America.

schizonte   A subgroup of protozoa (sporozoa) that reproduces asexually. Plasmodium is a schizonte that causes malaria.

Schwann cells   A type of brain cell that encompasses the axon of a neuron, thereby forming a sheath of myelin around the axon. The myelin sheath is essential for proper transmission of nerve impulses between neurons. Multiple sclerosis is an example of a disease that induces loss of muscle control by causing demyelination of the axon.

scintillation counter   A sensitive device for detecting single emissions of particles produced by radioactive decay.

secondary culture   The cell culture derived from the original outgrowth of cells derived directly from a tissue specimen, namely the primary culture.

secondary structure   The manner in which a linear polypeptide is folded, twisted, or otherwise bent. The most common types of secondary structure are the α-helix and the pleated-sheet structures.

second-order kinetics (bimolecular kinetics) A term describing the rate at which a chemical reaction involving two reacting molecules occurs.

segment polarity mutants Mutants of the fruit fly, *Drosophila melanogaster,* in which one of the halves of each segment (the P compartment) is replaced by the other half (the A compartment) so that each segment contains two mirror images of one of the normal halves.

segments, segmentation A pattern that develops in the embryo of the fruit fly, *Drosophila melanogaster,* defined by indentations giving the embryo the appearance of stacked discs with each disc representing a segment. Various structures of the adult, such as legs, antennae, wings, and eyes, develop from specific segments. Each segment consists of an A (anterior) compartment and a P (posterior) compartment.

selective medium A growth medium that, either by the inclusion of a toxic substance or by the lack of an essential nutrient, promotes the growth of only certain variant organisms in a population—for example the growth of penicillin-resistant bacteria on a nutrient agar containing penicillin. (See SYNTHETIC MEDIUM.)

self-assembly The spontaneous, unassisted assembly of the components of a complex structure, such as the protein viral coat of tobacco mosaic virus.

self-protein Any protein that, as the result of immunological screening in early life, is determined to be "self" and therefore not recognized as a foreign antigen that would be attacked by the immune system. Certain illnesses, referred to as autoimmune diseases, result from a failure of the immune system to recognize self-proteins.

self-tolerance The lack of an immune response to a self-protein.

semi conservative replication The mode of DNA replication in which each of the original parental DNA strands is based paired with one newly synthesized daughter strand. Experiments performed by Matthew Meselson and Franklin Stahl in the mid-1950s demonstrated that DNA replication was semi conservative as opposed to conservative. This finding laid the foundation for future experiments that ultimately elucidated the molecular details of DNA replication.

semidiscontinuous replication DNA replication involving the synthesis of many small fragments occurring on the lagging strand of double-stranded DNA in the form of OKAZAKI FRAGMENTS.

Sendai virus A member of the paramyxoviruses used to induce cell fusion, a technique for creating hybrid cells (heterokaryons) for the study of genetics in cultured cells.

sensitization A lowering of the threshold for a nerve impulse to be generated as a result of strong and repeated stimulation of a neuron by another neuron. Sensitization results from the tendency of some neurons to trigger an action potential if stimulated by weaker-than-normal nerve impulses or with shorter-than-normal refractory phases if action potentials have been triggered in that neuron in the recent past.

Sephadex A polysaccharide-derived gel (formed by cross-linking of dextran strands). Filtration of mixtures of biological molecules through Sephadex gels in columns is a widely used procedure for separation of molecules based on size. (See GEL EXCLUSION CHROMATOGRAPHY, GEL FILTRATION.)

Sepharose A form of agarose as small beads used in column chromatography for size separation of biomolecules. It is similar to Sephadex and serves as a matrix for attaching antibodies and other ligands for various types of affinity chromatography.

sequence   A term for the linear or end-to-end arrangement of biomolecules in a long polymeric molecule. Most often used to denote the order of purine and pyrimidine bases along the length of a nucleic acid, for example AAGCTTCTG. . . . , where A=adenine, C=cytosine, G=guanine, and T=thymine.

sequence conservation   The tendency of certain DNA sequences to resist change in the course of evolution and, therefore, to be similar in dissimilar organisms.

sequence homology   The degree of similarity between two nucleic acids as represented by the percentage of bases on one nucleic acid strand that match bases on the other nucleic acid strand when the two are aligned.

sequence tagged site (STS)   A means of cataloging sequence data by recording only that part of the whole sequence necessary to create PRIMERS that can be used to amplify the entire sequence from a DNA sample by the polymerase chain reaction (PCR).

sequencing   The process of determining the sequence of a polymeric biomolecule, such as a nucleic acid, or the amino acid sequence in a polypeptide, or the sequence of sugars in a polysaccharide. (See DIDEOXY SEQUENCING.)

serine   An amino acid that, because it contains an hydroxyl group, can serve as a site for phosphorylation when serine is part of a protein.

serodiagnostics   A diagnosis based on the indirect evidence provided by serology indicative of a disease state or that an individual has been previously exposed to a pathogenic organism (e.g., tuberculosis).

serologic reactions   Any of several reactions based on the presence of specific antibodies in the blood serum. These reactions generally fall into three categories: bacteriolysis, precipitation, and agglutination.

serology   A type of laboratory analysis based on the presence or absence of specific antibodies in the blood serum.

seropositive   **(– negative)**   The finding, in a diagnostic test, that reactive antibodies to a given agent are present (seropositive) or are not present (seronegative) in a sample of blood serum.

serotinin   A monoamine neurotransmitter made from the amino acid tryptophan.

serum   The liquid part of blood from which the blood cells have been removed by clotting.

serum albumin   One of the most abundant proteins in blood (albumin constitutes about 50% of the plasma protein). Albumin has at least two main functions: (1) regulate water content of the tissues, and (2) carry fatty acids in the bloodstream.

serum globulins   A group of abundant blood proteins alpha, beta, gamma with wide-ranging functions. The globulins are designated α, β, and γ. γ-globulins include all serum antibodies; the α- and β-globulins form essential complexes with various substances, such as lipids (these complexes are known as lipoproteins), carbohydrates (mucoproteins and glycoproteins), iron (transferrin), and copper (ceruloplasmin).

shadowing   The process of coating a specimen with a thin layer of metal, such as platinum or palladium, by heat evaporation under a vacuum. Shadowing is necessary to view surface detail of the specimen under an electron microscope.

shikimate pathway   A major biochemical pathway by which all the "aromatic" amino acids (tyrosine, phenylalanine, and tryptophan) are synthesized from one parent chemical, shikimate (an ester of shikimic acid).

Shine–Delgarno sequences   Special sequences present on the 5'-region of each gene in a prokaryotic cell. These sequences are rich in the bases adenine and guanine and help to align the ribosome on the mRNA in order for translation to begin at the proper start site.

shotgun cloning method   A technique of cloning a DNA sequence of interest based on mass LIGATION of a heterogeneous mixture of DNA fragments into a VECTOR; the vector carrying the DNA of interest is then selected from a mixture of cloned DNA fragments. This technique is useful when the DNA of interest is represented in low abundance or is difficult to purify.

shuttle vector   A vector genetically engineered to permit the growth and/or expression of recombinant DNAs in both prokaryotic and eukaryotic cells.

sialic acid   A modified sugar found in the lipids (of the membranes of neural cells) that are part of the receptor for neurotransmitters.

sickle cell anemia   A genetic condition involving a point mutation in the β chain of the hemoglobin protein, resulting in a loss of ability to carry oxygen from the lungs to the tissues of the body. The disease derives its name from the fact that red blood cells carrying the mutant hemoglobin assume an elongated sickle shape.

sickle cell disease   The pathological condition caused by sickle cell anemia characterized by an inability to handle exertion.

sigma factor   A small protein that forms a complex with the RNA polymerase enzyme in prokaryotic cells. The formation of sigma factor–RNA polymerase is essential for accurate intiation of transcription in bacteria.

signal peptidase   An enzyme that catalyzes the cleavage of the signal peptide immediately after the polypeptide is inserted into the ENDOPLASMIC RETICULUM.

signal-recognition particle (SRP)   A ribonucleoprotein, comprising six polypeptides and a small (7S) RNA molecule, that mediates the binding of a signal sequence on a preprotein to its appropriate membrane receptor.

signal sequence   A special sequence of amino acids on the amino-terminal end of polypeptides destined to be exported from a eukaryotic cell. If the signal sequence is present, the protein bearing that sequence is transferred into the endoplasmic reticulum where it is further processed for export.

silencers   Certain nucleotide sequences that act to suppress the activity of a PROMOTER. Silencers may act at distances greater than 1 kilobase away from the promoter sequences they act on.

silent mutation   A mutation whose effect is not manifest, either because it occurs in a nonessential region of DNA or because the effect of the mutation is masked.

silent sites   DNA nucleotide bases that, when changed, for example, by mutation, do not result in any change in the amino acids in the polypeptide coded for by the DNA.

simian virus 40   A small DNA virus accidentally discovered as a contaminant in cultured African green monkey kidney cells used to grow poliovirus for vaccine development. The virus was later found to be oncogenic in mouse but not in humans or in its natural host.

simple-sequence DNA   DNA sequences, generally very short, that are extremely highly repeated throughout the genome of an organism.

sindbis virus A member of the family of RNA-containing togaviruses (alphavirus group). Infection in humans and other mammals is via mosquito, and varying degrees of encephalopathy (brain disease) are produced.

sis oncogene An oncogene found in simian sarcoma virus and associated with sarcoma tumors in monkeys and cats. The name derives from Simian Sarcoma. The sis oncogene protein is virtually identical to one of the subunits of platelet-derived growth factor (PDGF).

sister chromatid exchange The exchange of material between the two daughter strands of a replicated chromosome (i.e., chromatids) during MITOSIS; recombination occurring at the chromosomal level.

site-directed mutagenesis Technique by which specific bases on a segment of DNA are experimentally altered.

site-specific drug delivery A technique, using various strategies for targeting drugs to certain tissues: For example, chemical linkage of antibodies to the drug molecule or attachment of the drug molecule to a ligand specific for a cell surface receptor. (See FUSOGENIC VESICLES).

site-specific recombination Recombination between two DNAs occurring at a specific site on each DNA. Site-specific recombination is exemplified by integration of λ BACTERIOPHAGE DNA in which recombination takes place at a site designated as attP on the bacteriophage DNA and the corresponding site (designated attB) on the bacterial host DNA.

skeletal muscle The relatively more striated muscle tissue associated with voluntary movement, for example, in the movement of the limbs.

small nuclear RNA (snRNA) A very short piece of RNA that complexes with a set of proteins (snRNPs) to form a structure whose function is to clip out loops in other RNAs, particularly for splicing RNAs destined to become mRNAs.

smooth ER The ENDOPLASMIC RETICULUM not bound to RIBOSOMES.

smooth muscle The relatively less striated (i.e., "smooth") muscle associated with involuntary movement (e.g., heart muscle).

S1 nuclease An enzyme that catalyzes the breakdown of any single-stranded nucleic acid or single-stranded region of a nucleic acid.

S1 nuclease mapping A technique of determining, on a segment of DNA, the precise location of the sequences from which a given RNA is transcribed.

sodium dodecyl sulfate (SDS) A detergent widely used to dissociate biological materials into their component molecules.

sodium dodecylsulfate–polyacrylamide gel electrophoresis (PAGE) A variation of the polyacrylamide gel electrophoresis technique in which SDS is dissolved in the polyacrylamide gel. This type of gel is widely used to separate proteins in mixture from one another on the basis of size.

sodium-potassium pump A specialized transmembrane protein that pumps sodium ions out of the interior of the cell while pumping potassium ions into the cell. Although sodium–potassium pumps are found in a variety of cell types, they are especially abundant in nerve cells where they serve to establish an electric potential across the membrane, which is the basis of nerve impulse transmission.

soma A term for the entire body of an organism without the reproductive cells.

somatic cell A nonreproductive cell; any cell that does not generate either sperm or egg.

somatic cell hybrid   The product formed by somatic cell HYBRIDIZATION.

somatic cell hybridization   Combining the genetic material of two cells by CELL FUSION, such as that induced by SENDAI VIRUS or polyethylene glycol (PEG).

somatic cell therapy   A gene therapy based on the introduction of new genetic material or alteration of existing genetic material in cells other than those giving rise to sperm or egg; for example, the introduction of insulin genes into pancreatic cells.

somatic mutation   Any mutation not affecting the reproductive cells. This type of mutation usually affects a particular tissue type and is not passed down to offspring in the form of a transmissible genetic defect.

somatomedin   A polypeptide hormone, produced in the liver, that induces growth of bone and muscle.

somatostatin   A polypeptide hormone, produced by the hypothalamus, that helps to regulate blood sugar levels by inhibiting the release of glucagon and insulin by the pancreas.

somatotropin   A polypeptide hormone, produced in the anterior pituitary, that simulates the liver to secrete somatomedin-1.

sorbitol   An alcohol derived from glucose. In diabetes, sorbitol accumulates in the eye, kidney, and other tissues leading to osmotic swelling and eventual damage of critical cells, such as the optic nerve.

SOS repair system   A system of at least 15 different proteins that work to repair severe DNA damage in bacteria; the system appears to be induced by the presence of an excessive amount of single-stranded DNA as might be generated by DNA damage.

Southern, E. M. (b.1918)   The discoverer of the Southern DNA blot hybridization technique.

Southern blot hybridization   Hybridization in a complex mixture of DNA fragments separated by size on an agarose gel; a technique for identification of a DNA fragment(s) by first transferring the DNA fragments from the agarose gel to a special membrane and then hybridizing the DNA fragments to a specific probe.

spacer DNA   Stretches of nontranscribed DNA separating transcribed regions of DNA that code for ribosomal RNAs.

species   Different forms of an organism among the members of a genus incapable of producing offspring by interbreeding.

specific activity   The activity present in some amount of a substance, as defined for that substance by convention; for example, units of enzyme activity per microgram of protein, units of hormone activity per milliliter of solution, distintegrations per minute per mole of ^{14}C-labeled amino acid.

sperm (spermatozoa)   The mature cell derived from the male reproductive cells (gametes) produced by MEIOSIS.

spermatids   Immature sperm cells having the HAPLOID number of chromosomes but lacking the morphological features of sperm, such as the elongated acrosome-bearing head and the tail assembly that make spermi motile.

spermatocytes   Cells representing stages in the formation of sperm: primary spermatocytes are cells containing the DIPLOID number of chromosomes but which, after dividing, form secondary spermatocytes containing the haploid number of chromosomes. The secondary spermatocytes differentiate to form spermatids.

S phase A part of the cell cycle during which the total COMPLEMENT of a cell's DNA is replicated.

sphingolipid A type of membrane lipid derived from the compound, sphingosine. Sphingolipids are subdivided into sphingomyelins, gangliosides, and cerebrosides, all of which are important components of the brain cell membranes. Altered metabolism of sphingolipids is the cause of the genetic syndrome Tay–Sachs disease.

spindle apparatus The bundles of microtubules attached at one end to the centromere of chromosome and at the other to the centriole and responsible for the movements leading to segregation of the chromosomes during cell division. (See MITOTIC APPARATUS.)

spleen A large ductless organ in the upper left portion of the stomach that plays a role in the maturation and differentiation of antibody-forming blood cells.

splice, splicing Joining of separated sections of an RNA molecule to generate new RNAs. In the process by which mRNAs are created from long RNA precursors in the nucleus, sections of RNA representing EXONS are spliced out so that segments representing exons are joined. Splicing is part of the process of RNA processing taking place in the nucleus. (See SPLICEOSOME, SPLICING JUNCTION.)

spliceosome A complex that mediates the splicing of an RNA molecule during mRNA formation. The spliceosome contains the RNA precursor in which the ends of the regions to be

RNA Transcript Splicing

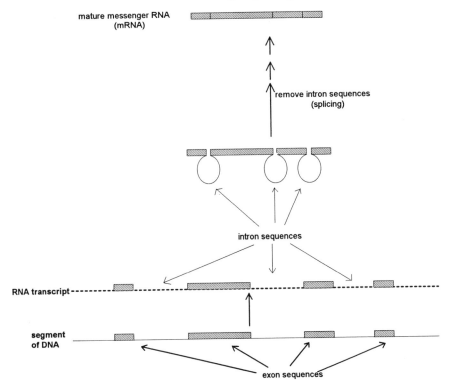

mature messenger RNA (mRNA)

remove intron sequences (splicing)

intron sequences

RNA transcript

segment of DNA

exon sequences

Spliceosome

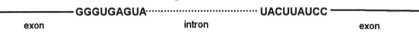

———————— GGGUGAGUA ·································· UACUUAUCC ————————

exon   intron   exon

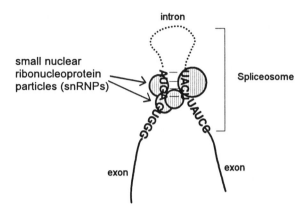

small nuclear ribonucleoprotein particles (snRNPs)

intron

Spliceosome

exon   exon

joined are held in place by small ribonucleoprotein particles (snRNPs).

splicing junction The site on a spliced RNA where the ends of the spliced RNA segments meet.

spontaneous mutation A change in a nucleotide base in the DNA occurring during the normal process of DNA replication and without the action of mutagenic agents.

src The oncogene carried by the Rous sarcoma virus, which produces sarcomas in birds. The product of the *src* gene is a phosphorylated protein denoted pp60. The src protein is a tyrosine kinase and is believed to cause transformation as a result of its ability to carry out phosphorylation of critical proteins.

staggered cut A term applied to the type of cleavage of DNA molecules produced by most restriction enzymes in which one end (usually the 5'-end) protrudes past the cut end of the other strand.

starch A complex polysaccharide used by plants as a means of storing glucose. Starch consists of long polymers of glucose joined to one another to form a compact, branched macromolecule similar to glycogen.

stationary phase The point at which, in a bacterial culture, the cells become so numerous that the nutrient supply is exhausted and growth ceases. (See GROWTH PHASES.)

stem cell Any cell that, in a tissue, is itself immature but gives rise, through cell division, to cells that become the mature form of the cells characterizing the tissue. The marrow in bone is a classic example of stem cells that give rise to the mature differentiated blood cells, including red blood cells, macrophages, and the antibody-producing cells of the immune system.

stem–loop structure A structure formed by nucleic acids, but particularly RNAs, in which a segment of the nucleic acid strand forms base pairs with a distant complementary sequence; the base-paired sequences form the stem, and the sequences intervening between the base-paired regions form the loop.

stereoisomer A form of a molecule involving different arrangements of atoms or molecules around a central atom, usually carbon in biomolecules. Stereoisomers are also referred to as optical isomers because crystals of stereoisomers cause polarized light to rotate in ways characteristic for each stereoisomer. (See DEXTROROTATORY ISOMER, LEVOROTATORY ISOMER.)

sterile   Completely free of living material.

sterilization   The process by which objects or liquids are made sterile, usually for the purpose of preventing disease, infection, or contamination. Common methods of sterilization include heating to temperatures above 125°C or prolonged exposure to ultraviolet light.

steroid   A class of potent hormones derived from cholesterol. Cortisone and the sex hormones estrogen and testosterone are examples of steroid hormones.

sticky ends   The single-stranded ends of any two nucleic acids whose nucleotide-base sequences are complementary to one another.

stimulatory neuron   A neuron that, when stimulated, functions to enhance the effects of the nerve impulse from another neuron.

stock culture   A culture of cells serving as a common source of cells for experimental purposes.

stop codon   A sequence of three nucleotide bases that do not represent the code for an amino acid but serve as signals for the termination of translation by the RIBOSOME. The three RNA stop codons are UAA, UGA, UAG.

stop transfer signal   For preproteins to be inserted into, but not completely through, a membrane, a stop transfer signal is a group of amino acids on the polypeptide that serves as a signal to stop its movement when the polypeptide is properly positioned in the membrane.

strand displacement   A variant of the normal mechanism of DNA replica-

Strand Displacement

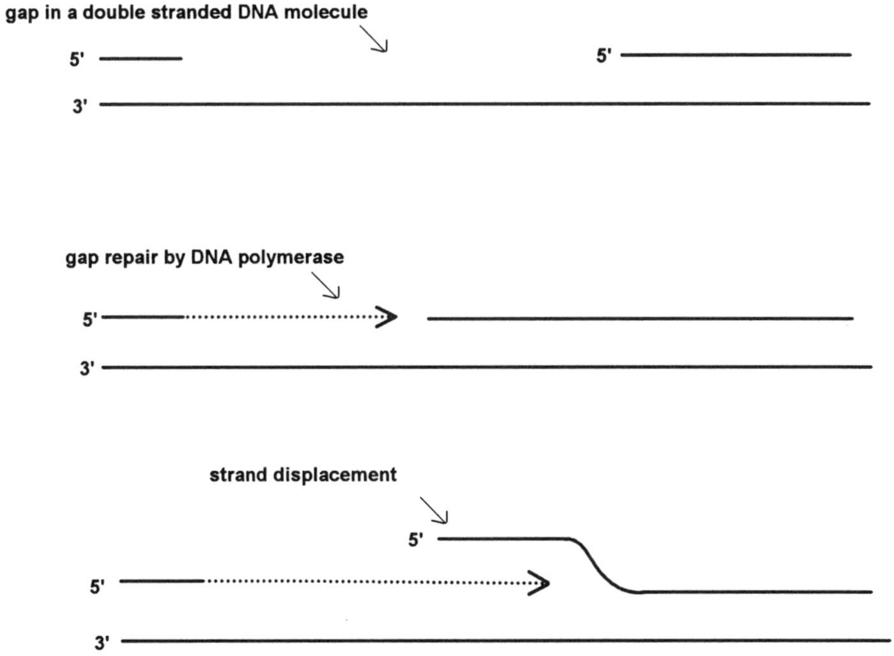

gap in a double stranded DNA molecule

gap repair by DNA polymerase

strand displacement

tion in which replication of one of the DNA strands proceeds from opposite ends of a linear DNA molecule.

streptomycetes   A funguslike bacterium found in soil. In addition to streptomycin, various isolates of streptomycetes have yielded more than 500 compounds of therapeutic value, including more than 90% of the clinically useful antibiotics.

streptomycin   An antibiotic, derived from molds, that exerts its antibacterial effect by causing bacterial RIBOSOMES to misread the CODONS on the mRNA, particularly with respect to the pyrimidines U and C, where one is usually mistaken for the other.

stress fibers   The fibrillar arrays, seen on the surface of a cell, oriented parallel to the direction in which a cell is moving. Stress fibers appear to be directly related to cell movement in that they are known to coincide with the actin filaments.

stringent response   A bacterial response to conditions of nutritional deprivation in which expression of nonessential genes is shut down. A stringent response involves a rapid downregulating of certain bacterial biosynthetic pathways (e.g., synthesis of ribosomal and transfer RNAs) when the amino acid supply becomes limited. (See ALARMONES.)

stroma   (1) The space between the GRANA and the CHLOROPLAST membrane containing some of the enzymes of the dark reaction in photosynthesis as well as the chloroplast RNA and DNA. (2) The connective tissue underlying the epithelial cell layer, for example, in skin, the digestive tract, and the airways in lung.

structural gene   A gene in an operon coding for the functional protein essential to the metabolism of the bacterial cell (e.g., an enzyme); as distinguished from the genes for a repressor protein that control the expression of a structural gene.

subcutaneous   Just underneath the skin; as in subcutaneous injections.

substrate   Any one of the reacting chemicals in an enzyme-catalyzed reaction.

substrate analog   A chemical similar in form to a particular substrate but that does not participate in the chemical reaction of the substrate. Substrate analogs are used for various purposes, such as inhibition of certain enzyme systems or for studying the mechanism of enzyme action. (See COMPETITIVE INHIBITION.)

subtilisin   A proteolytic enzyme (protease) produced by the soil bacterium *Bacillus amyloliquefaciens.*

subunit   One part of a complex biological molecule, such as an enzyme or ribosome. The subunits combined constitute the biologically active molecule.

sucrose   A disaccharide consisting of one molecule of fructose linked to one molecule of glucose. Common table sugar is sucrose.

sucrose density centrifugation   A technique that separates molecules in a mixture according to their density by using sufficiently high centrifugal force to cause the molecules to migrate through a solution of sucrose. In density gradient separation the sucrose solution increases in density the farther the molecules travel.

sudden-correction   model   The model that proposes that in gene clusters in which there are multiple copies of a gene (e.g., the genes coding for ribosomal RNA) the entire gene cluster is replaced "every so often" by a process that replicates the entire gene cluster from just one or a few copies. The sudden-correction model is actually an error-correcting mechanism that accounts for the fact that mutational errors in some of the gene copies do not accumulate over time.

sugar   Any compound that conforms to   the   general   molecular   formula

Straight Chain and Cyclized Forms of Sugars

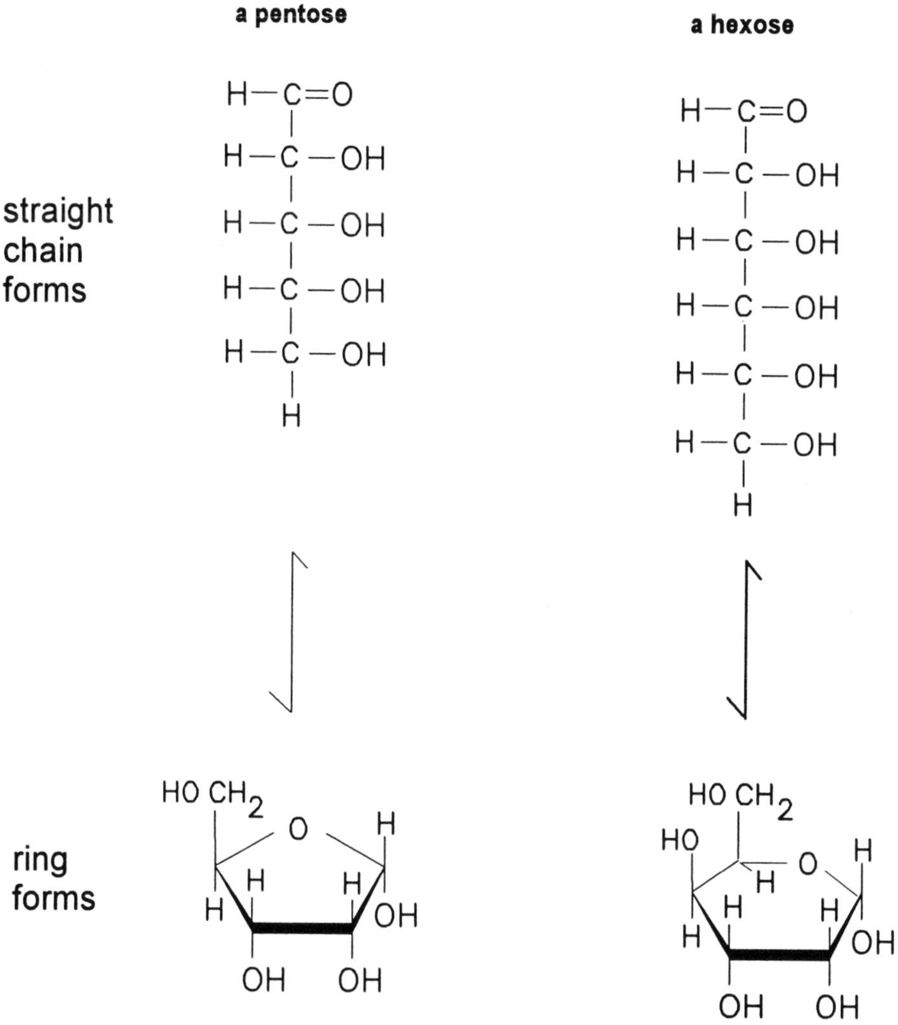

a pentose

a hexose

**straight
chain
forms**

```
H—C=O
  |
H—C—OH
  |
H—C—OH
  |
H—C—OH
  |
H—C—OH
  |
  H
```

```
H—C=O
  |
H—C—OH
  |
H—C—OH
  |
H—C—OH
  |
H—C—OH
  |
H—C—OH
  |
  H
```

**ring
forms**

$C_n(H_2O)_n$, where n is any number between 3 and 7 and where the carbon atoms are linked in a chain.

supercoiled DNA   A circular double-stranded DNA molecule in which the circle is itself twisted into a compact knot. This is the replicative form of many viral DNAs in their host cells.

suppressor gene   Any gene that acts to suppress the effects of mutation.

suppressor mutation   Any mutation that suppresses the effects of a previous mutation, such as a mutation that suppresses the effects of FRAMESHIFT MU-

Supercoiling of Circular DNA

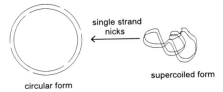

single strand
nicks

circular form

supercoiled form

TATION by reinstating the proper reading frame.

suppressor T cell   A type of T-lymphocyte that suppresses the antigenic response of antibody-forming T and B cells; that is, it inhibits the formation of antibody to a particular antigen.

suppressor tRNA   A mutation in a transfer RNA that suppresses the effect of a previous mutation in a gene. The suppressor mutation allows the suppressor tRNA to read the first mutation correctly, thereby ensuring the process of translation. (See AMBER SUPPRESSOR.)

surfactant   Any agent that lowers the surface tension of water. Soaps and detergents are the most common surfactants.

Svedberg unit   A measure of molecular size based on the rate of sedimentation of a molecule in a centrifugal field. The Svedberg unit is designated "S" and is not directly proportional to size; for example, s values of nucleic acids vary with secondary structure, temperature, and salt concentration.

symbiosis   A state of two or more organisms living in permanent close proximity for the mutual purpose of supplying some essential nutrient or life function to one another.

synapse   The specialized junction between the tip of the axon from a neuron and the dendrite of an adjacent neuron. The transmission of nerve impulses from one neuron to the next is carried out by neurotransmitters that cross the synapse.

synapsis   A stage in the recombination process mediated by the recA protein in which the recA protein forms a complex with the single-stranded and double-stranded DNAs that will then align with each other before undergoing recombination.

synaptic cleft   The space intervening between the axon and dendrite membranes in a synapse.

synaptic vesicle   A membrane-enclosed vesicle carrying the NEUROTRANSMITTERS to the synapse where they are released by fusion of the synaptic vesicle membrane with the membrane at the axon terminus.

synaptonemal complex   The structure joining chromosome pairs when homologous chromosomes align during the process of MEIOSIS.

synchronous culture   A cell culture in which all cells are simultaneously at the same phase of the cell cycle. Experimentally, synchronization of cells can be achieved by techniques that transiently, but specifically, block in one phase of the cell cycle, such as MITOSIS, leading to accumulation of cells at the block; synchronous growth ensues when the block is released.

syncitium   A multinucleated cytoplasm such as occurs when cells are fused by treatment with polyethylene glycol or Sendai virus.

synergism   Facilitation of a response by multiple stimuli such that the magnitude of the response is greater than the sum of the individual stimuli. The principle of synergism is often exemplified by the facilitation of a nerve impulse when a single neuron is stimulated by several excitatory neurons.

synthesis   The process of creating a new substance from precursor molecules. Biosynthesis is the process by which living cells create new biomolecules, whereas the term synthesis is generally applied to processes used in the laboratory for creating biomolecules.

synthetic medium   Solutions of nutrients created for the purposes of growing cells of various types in culture. Most synthetic media formulations attempt to recapitulate as nearly as possible the natural nutrient environment for the cell type being cultured.

synthetic peptides The creation of peptides in the laboratory using techniques of organic chemistry to link amino acids according to some prescribed sequence so that a peptide of any given primary structure can be synthesized.

syntrophism Crossfeeding by organisms sharing a common growth medium, such as bacterial colonies whose growth depends on a factor or factors secreted by a neighboring bacterial colony on a common agar plate.

syphilis A venereal disease caused by the spirochete *Treponema pallidum*. If left untreated the disease may cause blindness and neurological symptoms, including a syndrome characterized by a loss of motor control known as general paresis.

systemic lupus erythematosus An autoimmune disease of the connective tissue characterized by a reddish skin rash (erythema) and a wide variety of conditions related to internal organ malfunction. Antibodies to a wide variety of self-antigens are seen.

T

tandem In general, a group of objects arrayed in a line, one next to the other. As applied to molecular genetics, the term refers to genes arranged in tandem along a stretch of DNA. A number of viral and cellular genes (e.g., rRNA genes) that undergo amplification are tandemly arrayed.

T antigen(s) The products of the early genes of the papova viruses. In normal hosts the T antigen(s) function to stimulate viral-DNA replication and to regulate expression of the viral genes. However, in host cells that do not support virus replication, the T antigens are responsible for transforming the cells to a cancerous PHENOTYPE.

TATA box Another name for the Pribnow box.

tautomerism The rapid and continual transition between different forms of a molecule based on delocalization of an electron(s) on different atoms of the molecule.

taxol A plant alkaloid that stabilizes microtubules, thereby freezing cells in mitosis. Since cancer cells are rapidly dividing cells, taxol is currently being considered as a chemotherapeutic agent. (See MITOSIS.)

taxonomy The science of classification, usually referring to plants and animals.

Tay–Sachs disease A hereditary disease in which accumulation of a certain type of sphingolipid accumulates in the brain and spleen, leading to degeneration of the nervous system and death at an early age.

T cell A LYMPHOCYTE named for the thymus (i.e., "thymus" cell) where the majority of T cells mature. T cells are responsible for so-called cell-mediated immunity, the immune function directed toward detecting and destroying foreign cells rather than foreign proteins.

T-DNA   A term for the Ti PLASMID carried by *Agrobacterium,* a parasite that induces various plant tumors. The tumors are a direct result of expression of the genes carried by Ti in the plant cells.

teichoic acids   Long polymers of glycerol or ribitol molecules linked by phosphate groups. Teichoic acids are a structural component of the outer cell wall of gram positive bacteria. (See GRAM STAIN).

telomere   The terminal portion of a chromosome.

telomeric sequences   Special sequences on the ends of DNA strands required for synthesis of the terminal segments of the lagging strand. Telomeric sequences are present on the ends of chromosomes (the telomeric region) and are used in the construction of yeast artificial chromosomes (YACs).

telophase   The stage of MITOSIS in which the new cell membrane dividing the daughter cells forms (the cell plate) and chromosomes reform into diffuse chromatin.

temperate phage   A BACTERIO-PHAGE capable of establishing lysogeny in a host rather than undergoing a normal LYTIC CYCLE.

temperature-sensitive mutant (Ts mutant)   Any organism expressing a function with a temperature dependence; for example, bacteriophages that establish lysogeny at one temperature but not at another.

template   In general, a pattern for creating a copy of something. A nucleic acid strand whose sequence is used to create a complementary nucleic acid copy.

teratoma   A type of tumor derived from a developing embryo.

terminal redundancy   During the replication of certain bacteriophage such

Terminal Redundancy

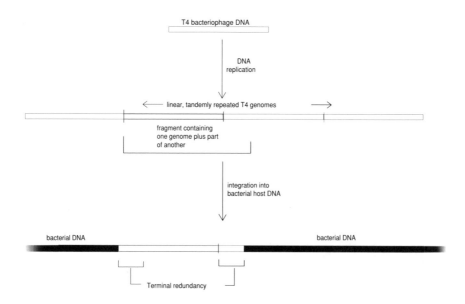

as T4, slighty more than one GENOME equivalent is cut from a long linear DNA representing T4 genomes repeated in end-to-end, head-to-tail fashion, leading to a genome segment in which the ends are repeated.

terminal transferase An enzyme that catalyzes the addition of an unspecified number of deoxyribonucleotides from deoxyribonucleotide triphosphates (dNTPs) to a free 3'-hydroxyl end of double- or single-stranded DNA:

$$3'\text{OH}\text{——}5' \quad \begin{array}{c} \text{terminal} \\ \text{transferase} \\ + \text{dNTPs} \\ \longrightarrow \end{array} \quad \text{NNNNNNNNNNN}\text{——}5'$$

Terminal transferase catalyzes the addition to DNA of long polymeric chains of whichever nucleotide whose triphosphate (dATP, dCTP, dGTP, or TTP) is used as a substrate. The addition of long nucleotide tails to DNA fragments is a tool for cloning DNA fragments into VECTORS for which no convenient RESTRICTION ENZYME sites exist.

ternary initiation complex A three-part complex necessary to start the process of translation. The ternary initiation complex consists of met-tRNA, GTP, and an initiation factor (eIF2).

tertiary structure The overall, three-dimensional folding of a polypeptide; the folding, twisting, or conformation of the secondary structure of the polypeptide.

testosterone The steroid hormone produced by the testes that regulates sperm production and male sexual behavior.

tetanus A syndrome caused by infection by the anaerobic bacterium *Clostridium tetani*. The disease symptoms (uncontrollable muscle spasms) are due to the presence of a potent neurotoxin produced by the bacterium.

tetracycline A broad-spectrum (both gram-positive and gram-negative) antibiotic produced by *Streptomyces venezuelae*.

12-O-tetradecanoylphorbol-13-acetate (TPA) A plant-derived phorbol ester that is a potent tumor promoter. TPA appears to exert its tumor-promoting activity by activating protein kinase C in the cell membrane.

thalassemia A disease resulting from a mutation, often a deletion of DNA within the gene, that causes a reduction or complete loss of expression of one or both of the globin proteins (α or β), resulting in gross defects in hemoglobin function that may be fatal. The thalassemias are examples of genetic disease brought about by uneven crossing over.

thalassemia, β A type of thalassemia affecting the biosynthesis of the β-globin chain of hemoglobin. Some β-thalassemias have been found to be the result of defects in gene regulation, such as RNA processing in the nucleus, and so have provided important insights into mechanisms of gene expression control.

thermoinducible Stimulated by heat. In the context of molecular genetics the term is usually applied to genes and/or their products whose activity is rapidly increased when the temperature rises by several degrees higher than optimum for the growth of the organism.

thermophile An organism that thrives at high temperature.

***Thermus aquaticus* (Taq) polymerase** A DNA polymerase isolated from the thermophilic bacterium *Thermus aquaticus*.

theta structure The term describing the structure formed when a circular double-stranded DNA molecule is engaged in replication that proceeds in

Theta Structure

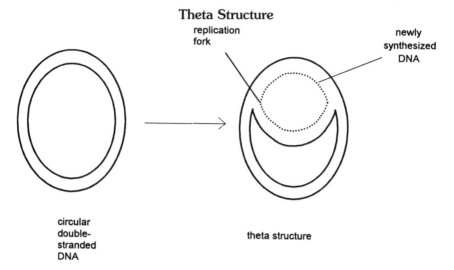

replication fork

newly synthesized DNA

circular double- stranded DNA

theta structure

both clockwise and counterclockwise directions from the same starting point.

thiamine   Also known as vitamin B_1. An important cofactor for the reactions involved in the O_2-dependent oxidation of sugars (respiration) in energy (ATP) production.

thin-layer chomatography   A sensitive analytical technique for separating molecules on the basis of their differing solubility in various solvents. The sensitivity of the technique derives from running the sample on a thin inert matrix to maintain the sample in a concentrated form.

6-thioguanine   A purine derivative acted upon by the enzyme hypoxanthine-guanine phosphoribosyl transferase (HGPRT) to form a toxic compound. For this reason 6-thioguanine is used to select cells containing low levels of HGPRT. (See HYPOXANTHINE-AMINOPTERIN-THYMINE SELECTION.)

thiol   The sulfur-containing analog of an alcohol (—OH) group.

thiostrepton   An antibiotic that acts by blocking a critical step (translocation) in protein synthesis (i.e., translation) by binding to the large subunit of the ribosome.

1-thiouridine   An unusual pyrimidine base found only in TRANSFER RNA (tRNA). Thiouridine is derived from uridine by the replacement of an oxygen atom with sulfur.

threonine   An amino acid that, like serine and tyrosine, contains an —OH group on its side chain. For this reason threonine serves as a site of phosphorylation in proteins.

thrombin   In the blood-clotting pathway, the enzyme that produces fibrin from fibrinogen by cleavage of a portion of the fibrinogen molecule. Fibrin polymerizes to form a clot.

Thy-1   A protein on the surface of T CELLS with homology to a portion of the Fc region of immunoglobulins. Like other such T-cell membrane proteins, Thy-1 is believed to play a role in the recognition of foreign antigens.

thylakoid disk   A membrane-enclosed, coin-shaped structure containing the light-reacting pigments and other components of the light reaction in photosynthesis; these components are contained in the membrane of the thylakoid disk.

thymectomy   Removal of the thymus by surgery. A procedure frequently performed to render an animal unable to mount an immune response to foreign cells, such as tissue grafts or lymphocytes from another animal.

thymidine   A pyrimidine base attached to the deoxyribose sugar in deoxyribonucleotides.

thymidine kinase   An enzyme that catalyzes the major step in the formation of TTP from thymine. Since this pathway is the only means by which thymine or thymine analogs can enter nucleic acids, manipulation of this enzymatic step provides an important experimental tool for studying gene action by the incorporation of modified bases into DNA.

thymidine triphosphate (TTP)   The thymine-containing nucleotide precursor of DNA.

thymine   One of the nitrogenous bases found in nucleic acids. Thymine is a PYRIMIDINE which forms hydrogen bonds with the purine ADENINE.

thymine dimers   A type of mutation in which adjacent thymine bases in a DNA strand are covalently linked to one another, causing the two bases to be read as a single base during DNA replication or transcription. Thymine dimers

Thymine

are largely caused by exposure of tissue to ultraviolet light. (See ULTRAVIOLET (UV) REPAIR.)

thymosin   A mixture of small naturally occurring peptides that acts to promote the appearance of the T-cell surface proteins seen in mature T cells (e.g., Thy-1). Thymosin is thought to mimic the effects of the hormone(s) normally inducing maturation of prothymocytes.

thymus   A structure comprising lymphatic tissue located in the upper portion of the chest cavity in mammals. Some of the immature lymphocytes from the bone marrow migrate to the thymus where they develop into the mature lymphocytes responsible for cellular immunity. In mammals the thymus is present in young animals but decreases in size or disappears in adults.

thyroxine (T4)   One of two major hormones secreted by the thyroid gland in response to thyrotropin, a pituitary hormone. Thyroxine is made from iodinated tyrosine and has the effect of raising the basal metabolic rate, an indicator of the oxygen-dependent oxidation of sugars.

tight junction   A structure in which the cell membranes of neighboring epithelial cells are brought into extremely close contact, preventing the seepage of even small molecules through the space between cells. Tight junctions are particularly evident between the epithelial cells lining the gut, where they function as a barrier against unregulated diffusion of substances into the bloodstream from the digestive tract.

tissue culture   The general technique of keeping tissues and/or cells derived from tissues alive outside the organism from which they were derived by creating an artificial environment providing the essential aspects of the natural setting. The development of sophisticated tissue culture systems has been a major factor in recent advances in biomedicine since tissue culture permits organ-specific cells to be studied and manipulated in an experimental environment.

tissue plasminogen activator (tPA) An enzyme that catalyzes the cleavage of the blood protein plasminogen to the active form of the blood-clotting protein plasmin. tPA has recently been used as a therapeutic agent for destroying blood clots in the blood vessels. The tPA gene has been cloned, and the protein has been synthesized in large quantities using recombinant-DNA techniques, so it may find widespread therapeutic use as a preventative agent for heart attack and stroke.

titer The concentration of live virus in a fluid; the number of plaque-forming units (PFU) or focus-forming units (FFU) per unit volume of fluid (e.g. PFU/mL).

titration The process of experimentally determining the concentrations of substances by adding known amounts of chemical antagonists to the solution until the effects of the target substance are neutralized.

Tn5 A type of insertion sequence that carries the gene for resistance to the antibiotic kanamycin. (See TRANSPOSON.)

Tn10 A type of insertion sequence that carries the gene for resistance to the antibiotic tetracycline. (See TRANSPOSON.)

tobacco mosaic virus (TMV) A large, filamentous, RNA-containing plant virus. TMV was one of the first viruses to be studied in detail; among the findings derived from studies on TMV was the spontaneous assembly of the viral coat from its component subunits.

tonofilament A filament type characteristic of epithelial cells. Tonofilaments are approximately 8–10 nm in diameter by transmission electron microscopy and terminate as filament bundles at cell junctions characteristic of epithelial cells, known as desmosomes. Tonofilaments have been shown to be identical to keratin filaments making up the INTERMEDIATE FILAMENT network characteristic of epithelial cells. (See INTERMEDIATE FILAMENT.)

topoisomerase A class of enzymes that catalyze the relaxation of supercoiled DNA by creating transient nicks in the DNA strands permitting tightly wound DNA to uncoil.

totipotent The concept that a particular cell (e.g., the fertilized egg) has the capability to generate, or differentiate into, any cell type in the body of an organism. Because the DNA in all cells of the body was believed to be essentially equivalent, the concept of totipotency was originally thought to apply even to specialized cells of a highly differentiated structure, such as the eye, but modern understanding of the fluidity of the genome now suggests that, in many cases, differentiation is accompanied by alterations in the DNA.

toxin A chemical poison secreted by one organism for purposes of defense against a competing organism. Because toxins normally target highly specific cell-organ systems, many toxins have been used to gain insight into normal biochemical mechanisms; for example, tetrodotoxin, a toxin secreted by the puffer fish, specifically paralyzes the sodium transport channel in nerves and has been used to study ion transport in the neural system.

T7 promoter A sequence that forms the PROMOTER for transcription of the genes of the T7 BACTERIOPHAGE. The T7 promoter is widely used in synthetic cloning vectors where expression of recombinant DNAs is desired.

trans In general, on the opposite side of, or across from. In organic chemistry, the term refers to a molecular configuration where groups are on the opposite side of a chemical bond from one another. In molecular genetics, the term is used to indicate changes in expression of a particular gene caused by an agent acting on the gene from a distance (e.g., a hormone) and not by an agent located on the same DNA strand (e.g., a promoter or enhancer sequence).

trans acting   Pertaining to a genetic element exerting an effect on a target located on a physically separate unit. For example, a gene coding for a regulatory protein is said to be trans acting with respect to the genes it controls because the target genes may be located on DNA strands or even chromosomes at some distance from the regulatory gene.

transamination   A type of biochemical reaction that allows amino acids, and therefore proteins, to enter the same biochemical pathway by which sugars are oxidized for energy (i.e., ATP) production.

transcellular transport   A mechanism for carrying certain substances from one side of a cell to the other. This type of transport is the means by which substances (e.g., glucose) move across the epithelial cells lining the intestinal tract to the bloodstream.

transcription   The process of making an RNA complementary to a strand of DNA. In transcription an RNA polymerase, using the order of nucleotide bases present in the DNA template as a guide, assembles nucleotides from the four ribonucleotides (ATP, CTP, GTP, UTP) to create the RNA strand.

transducin   An alternative term for G protein.

transducing phage   A bacteriophage that during its normal replicative cycle occasionally packages some of the DNA from the host into the bacteriophage head along with the normal bacteriophage DNA. The DNA so packaged can then be carried from the previous host and introduced into a new host infected by the transducing BACTERIOPHAGE.

transduction   The term for the process of carrying sections of DNA from one bacterial cell to another by a transducing bacteriophage.

transfection   The technique of introducing DNA into eukaryotic cells. Transfection is the process homologous to transformation in bacteria. Transfection encompasses techniques utilizing different principles to introduce the DNA, including electroporation and precipitation by calcium phosphate.

transfer factor   An as yet unidentified factor extracted from living T cells that when taken from one human and injected into another, induces some of the cell-mediated immunity that was present in the donor.

transferase   A class of enzymes that catalyzes the transfer of a chemical group from one substrate to another, such as methyl transferases for transfer of methyl groups from one molecule to another.

transferrin   A plasma protein that carries iron in the blood. Transferrin transfers the bound iron to the appropriate cells via a special cell-surface receptor.

transfer RNA (tRNA)   A type of RNA that recognizes the codon on the mRNA during the process of translation and brings the proper amino acid (attached to one of the tRNA) into close proximity to the end of the peptide chain being synthesized so that the amino acid can be added to the peptide chain. The tRNA molecule is folded so that a group of three nucleotides complementary to the codon in the middle of the molecule (the anticodon) is exposed, and the end of the tRNA is used for attachment of the amino acid corresponding to the codon. (See ADAPTOR MOLECULES.)

transformation, cancerous or neoplastic   The process by which a normal cell comes to attain the characteristics of a cancerous cell. Since the actual process of transformation cannot be directly observed, steps in the process are inferred by the expression of certain properties that cells taken from tumors exhibit when grown in culture, such as the ability to grow without being attached to a solid surface, lowered dependence of growth on serum, and others.

transformation, DNA   The process of introducing foreign DNA into bacteria. (See COMPETENCE.)

transforming growth factor (TGF)   See SARCOMA GROWTH FACTOR.

transgenic animal   An animal, typically a goat, pig, cow, or horse, modified by introduction of a foreign gene into its germline so that some specific aspect of phenotype, such as production of a human protein in the milk or resistance to disease, is conferred on the offspring.

transit peptide   A preprotein destined for insertion into a MITOCHONDRION.

translation   The process of assembling amino acids to form a polypeptide; the sequence of amino acids is specified by the CODONs on the mRNA used as the template. Translation is carried out on RIBOSOMEs with sites for tRNAs carrying the appropriate amino acids.

translational domain   One of two major classes of binding sites on the ribosome. The factors that bind within the translational domain are directly involved in the translation of mRNA into proteins. The translational domain contains binding sites for peptidyl transferase, mRNA, EF-TU, EF-G, and 5S RNA. (See ELONGATION FACTORS.)

translocation   (1) During translation, the process of moving the tRNA carrying the growing polypeptide chain from one site on the ribosome to another to make room for an incoming tRNA carrying a new amino acid. (2) In biochemistry, the process of actively transporting a molecule across a membrane. (3) The breakage of a chromosome followed by subsequent rejoining of one of the pieces to another chromosome. (See RECIPROCAL TRANSLOCATION.)

transmembrane protein   A protein inserted into and spanning the cell membrane so that one end of the protein protrudes from the cell (the extracellular domain) while the other end (the internal domain) remains in the interior of the cell. The two major functions of transmembrane proteins are to serve (a) as channels for, or transporters of, specific molecules and (b) as devices for transmembrane signaling. (See INTEGRAL MEMBRANE PROTEIN.)

transmembrane signaling   A signal mechanism in which the binding of a specific molecule (the ligand) to the extracellular domain of a transmembrane protein (called the ligand binding domain) causes a physical change in the internal domain that then sets in motion a series of chemical reactions, such as phosphorylation(s) of certain proteins that produce specific changes in the cell behavior. This is called signaling since the ligand never actually enters the cell. (See G PROTEINS.)

transmission electron microscope (TEM)   A device, similar in principle to a conventional microscope, that uses an electron beam instead of light and a magnetic field instead of a glass lens to focus the beam on the specimen. The image of the specimen is seen as a pattern of greater or less electron intensity in the beam emerging from the specimen. The great advantage of the electron microscope over the conventional light microcope is that, since electrons have a much shorter wavelength than photons, resolution of much finer detail is possible.

Transposon Tn3

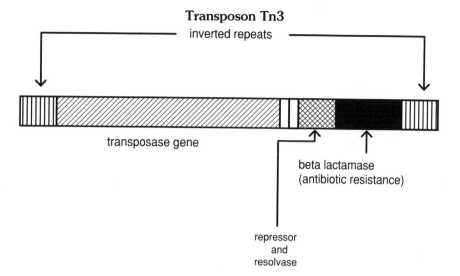

inverted repeats

transposase gene

beta lactamase
(antibiotic resistance)

repressor
and
resolvase

transplacement vector A VECTOR designed to transfer a defined segment of DNA of interest to another vector by recombination. Transplacement vectors are useful in situations where one wishes to introduce a DNA segment into a particularly large vector (e.g., baculovirus or a YAC) or other vector in which it is difficult to engineer a unique site for making the recombinant using conventional restriction enzyme technology.

transplant The removal of a tissue or portion of a tissue from its natural location and placement in a new location either in the same organism or in some other organism.

transplantation antigens Certain proteins, produced by the major histocompatibility locus, found on the surface of all cells in the animal and responsible for provoking rejection of tissue grafts. (See MAJOR HISTOCOMPATIBILITY COMPLEX.)

transport In biochemistry, the process of moving a molecule from one location to another, usually across a membrane. The term implicitly indicates that expenditure of energy is required for the transport. (See ACTIVE TRANSPORT.)

transport protein A transmembrane protein that mediates the transport of a molecule across a membrane. Transport proteins are often in the form of a channel spanning the membrane and allowing molecules to pass through. (See INTEGRAL MEMBRANE PROTEIN.)

transposase An enzyme, encoded by genes on a type of transposon, called an insertion sequence, that recognizes the terminal inverted repeat sequences and catalyzes the events in the transposition.

transposon A mobile, genetic element, found primarily in prokaryotic cells, that carries genetic information from one site in the genome to another. Transposons may carry a variety of genes or other genetic units but always carry the information needed for the transfer function, for example, short terminal-inverted repeat sequences needed for inserting the transposon into their target sites in the genome.

transposon yeast (TY) 1 elements A type of transposon in yeast; The Ty elements are about 6.3 kb long and contain genes similar to the GAG and POL genes of retroviruses.

trehalose   A disaccharide of the sugar glucose. Trehalose is mainly found in insects where it is used as an energy source.

tricarboxylic acid (TCA) cycle   A series of reactions beginning with the formation of citric acid (also known as the citric acid cycle or Krebs cycle) by which the majority of the energy from the oxygen-dependent oxidation of glucose is derived. The cycle consists of a critical series of reactions in the oxidation of sugars in which CO_2 and H_2O and electrons to be used in electron transport are produced from the intermediate acetyl CoA. The majority of CO_2 produced in the body from muscular activity is generated via this pathway.

trigger factor   A bacterial protein that, by forming a complex with a pre-protein, holds the polypeptide in a specific conformation necessary for its insertion into the bacterial cell membrane.

triglyceride   A type of lipid formed from glycerol and fatty acids. Triglycerides are the major form of storage for fatty acids and are the main constituents of body fat.

triplet   In general, any group of three. In molecular genetics, the term generally refers to the triplet of nucleotide bases making up the genetic code for an amino acid. (See CODON, WOBBLE HYPOTHESIS.)

triploidy   The state of having one extra haploid set of chromosomes—3n, where n=the haploid chromosome number.

triskelion   A subunit of the clathrin coat of coated pits. The triskelion is named for its tripartite structure consisting of three light and three heavy polypeptide chains.

Triton X-100   A nonionic detergent used for the biochemical isolation of cell membranes and nuclei.

T4 RNA ligase   An enzyme isolated from bacterial cells infected with the bacteriophage T4 that catalyzes the formation of a covalent bond between the phosphate group on the 5'-end of either single-stranded DNA or RNA and the 3'-hydroxyl end of either single-stranded DNA or RNA.

trypanosome   Any of a group of parasitic protozoa living in the blood of mammals and responsible for a variety of illnesses. Sleeping sickness, which is transmitted by the bite of the *tsetse* fly in Africa, is an example of a trypanosome-induced illness.

trypsin   A protease that cleaves polypeptides specifically after the amino acids arginine or lysine.

tryptophan   An amino acid containing an indole group in its side chain. Tryptophan is the precursor of the neurotransmitter serotonin in animals and the plant hormone indole acetic acid (IAA).

TSH-releasing hormone (TRH)   A short (three amino acids) polypeptide hormone, produced in the hypothalamus, that stimulates the anterior portion of the pituitary gland to secrete thyroid stimulating hormone (TSH).

tuberculosis (TB)   A chronic disease of the lungs caused by airborne infection by *Mycobacterium tuberculosis* and characterized by the formation of pulmonary lesions known as tubercules. If untreated, TB may ultimately cause necrosis throughout a large enough area of lung tissue that normal gas exchange in respiration is compromised and may be fatal.

tubulins Two proteins (α and β) of about 55,000 daltons each of which constitutes the subunit proteins of microtubules.

tumor An abnormal growth of a tissue. In benign tumors, the tissue growth eventually stops and the tumor remains confined to the site at which it began; in malignant tumors, growth is unlimited and may take place at sites distal from the site of origin, known as metastases.

tumorigenesis The process of tumor formation. Tumorigenesis is the end point in the transformation process.

tumor-inducing (Ti) plasmid See AGROBACTERIUM, CROWN GALL PLASMID.

tumor necrosis factor (TNF) A factor originally isolated from serum and found to be produced by certain leukocytes (e.g., macrophages) that appears to be selectively toxic to tumor cells. The gene for TNF has been cloned and expressed in quantities sufficient for ongoing clinical studies of its effectiveness as an anticancer agent.

tumor promoter A chemical that, by itself, does not cause tumors but will induce the formation of tumors after exposure of a tissue to any of a class of other agents called tumor initiators.

tumor virus Any virus that induces the formation of a tumor in the tissue it infects.

turgor/turgor pressure Water pressure resulting from the diffusion of water into cells. The term is usually applied to the rigidity of plant structures (e.g., leaves and stems) resulting from osmotic pressure.

Turner's syndrome A genetic defect in which the cells of the afflicted individual contain one X chromosome but no Y chromosome. Although such individuals have the outward appearance of females, the sexual organs are incomplete or undeveloped.

turnover number For an enzyme-catalyzed reaction, the number of substrate molecules undergoing reaction per unit time. For example, in the reaction catalyzed by carbonic anhydrase,

$$CO_2 + H_2O \xrightarrow{\text{carbonic anhydrase}} H_2CO_3$$

the enzyme has a turnover number of 600,000 molecules of CO_2 per second.

Tween 80 A nonionic detergent used for isolation of cell membranes and to reduce backround signal interference in immunoblotting.

twin sectors If the transposition process carried out by a transposon occurs just at the time of cell division such that only one daughter cell carries the transposition, then the descendents of such genotypically different daughter cells are called twin sectors.

twisting number (T) For a given double-helical DNA, a number representing the total number of helical turns. The twisting number is calculated as the total number of base pairs in the DNA under consideration divided by the number of base pairs per turn. (See LINKING NUMBER.)

tyrosine An amino acid with a phenolic group in its side chain. Tyrosine is a precursor of adrenaline and the skin pigment, melanin.

tyrosine kinase A type of enzyme that catalyzes the transfer of a phosphate group (phosphorylation) to a tyrosine amino acid in a protein. The oncogenes activated by a number of retroviruses have been shown to be tyrosine kinases. (See KINASE.)

U

ubiquitin A histonelike protein found in the nucleosomes of the fruit fly, *Drosophila*. Ubiquitin is thought to play a role in the condensation of chromosomes in MITOSIS.

ultracentrifugation Centrifugation at speeds great enough to produce forces of greater than 300,000 times the force of gravity. The forces produced by ultracentrifugation are capable of separating molecules of different sizes from one another. The ultracentrifuge was invented by the Swedish physical chemist Svedberg in 1923.

ultrafiltration A technique using gas pressure to drive samples through an ultrafine-meshed filter for the filtration of particles smaller than bacteria.

ultrastructure Features of cell architecture discerned in an electron microscope.

ultraviolet radiation Light with a wavelength in the range of about 200–400 nm. Ultraviolet light is readily absorbed by tissue exposed to it (e.g., skin). Ultraviolet radiation produces numerous alterations in biomolecules, including mutations in DNA linked to tumorigenesis. (See THYMINE DIMERS.)

ultraviolet repair A system used by bacteria to repair regions of DNA damaged by exposure to ultraviolet light. The uv repair system removes altered nucleotides (e.g., thymine dimers) and then recopies the region using the remaining strand as a template. (See EXCISION REPAIR.)

unequal crossing over A form of recombination in which a segment of DNA from one chromosome is transferred to a position adjacent to its own allele on the homologous chromosome:

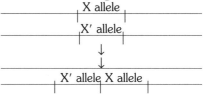

Uridine

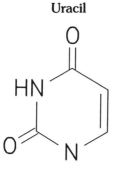

Uracil

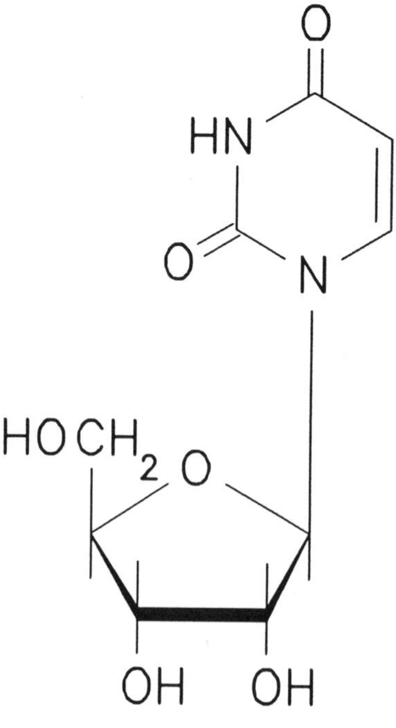

unicellular Having a single cell; for example, protozoa are unicellular organisms.

upstream activator sequences DNA sequences found in yeast that are similar to enhancer sequences in higher organisms in that they can stimulate the transcription of a gene from a long distance, but only when they are located upstream (i.e., in the 5'-direction) from the gene.

uracil A pyrimidine base used instead of thymine in ribonucleotides and RNA.

urea A compound derived from ammonia and carbon dioxide. Ammonia produced from the breakdown of amino acids is excreted as urea in the urine.

uric acid A purinelike molecule formed as a degradative product of the purine nucleotides. Because uric acid is relatively insoluble in body fluids, overproduction of it leads to gout, the accumulation of uric acid crystals in joints. Uric acid is the main vehicle for the elimination of ammonia by birds and reptiles.

uridine The nucleoside derivative of uracil—uracil bonded to ribose.

uridine triphosphate (UTP) The triphosphate derivative of uridine, the uracil-containing molecule used in the synthesis of RNA. (See NUCLEOTIDE.)

urokinase A protease found in urine and blood with the same activity as plasminogen activator; urokinase is therefore used therapeutically to dissolve blot clots that may accumulate in the coronary arteries.

V

vaccination The capacity of a vaccine to induce an immune response to a pathogenic organism in an individual, usually by multiple, direct injections of the vaccine at a body site favorable for exposure to the immune system. Vaccination is a prophylactic procedure generally ineffective in treating disease after onset.

vaccine A preparation derived from inactivated or attenuated pathogenic organisms used for vaccination.

vaccinia virus A member of the poxviruses that causes cytopathic destruction of epithelial cells (pocks); vaccinia is the agent responsible for cowpox. Vaccinia virus is closely related to variola virus, which causes smallpox in humans. Vaccinia is easily grown in tissue-culture cells for purposes of creating vaccines. For this reason vaccinia is being researched as a vector for expressing recombinant DNA for other pathogens with a view toward using this as a system for creating vaccines to other agents.

vacuole A membrane-enclosed cytoplasmic organelle generally arising from phagocytosis and often containing enzymes involved in degradation of biologic material (e.g., proteases).

valine An amino acid with an isopropyl group as its side chain; valine belongs to the group of nonpolar amino acids.

valinomycin An antibiotic comprising lactate, hydroxyisovalerate, and the amino acid valine, joined in a ring configuration that carries a potassium ion in its center. In this way valinomycin acts as a vehicle for transporting potassium through the cell membrane, thereby destroying the delicately balanced ion concentration in the cell. (See IONOPHORE.)

van der Waals forces   Weak electrostatic forces between nonpolar hydrocarbon molecules, such as those making up paraffins and the lipids in membranes. Van der Waals forces are primarily responsible for the aggregation of waxy, oily, or fatty substances in water.

variable region   The terminal portion of an antibody molecule containing the site that binds the antigen that the immunoglobulin carries specificity for. The variable nature of this region is a reflection of the many antigens for which specific antibodies can be made.

variable-surface   glycoprotein (VSG)   The major component of the cell surface of the trypanosome parasite such as the sleeping-sickness-producing parasites carried by the tsetse fly. The VSGs, the only antigenic molecule on the trypanosome surface, change throughout the development of the organism. These rapid changes in VSG allow the organism to evade the immune system of the host. The VSGs are coded for by a large family of genes, and the expression of different VSGs at various stages of development is due to the activation of different VSG genes from this gene family.

vasopressin   A polypeptide hormone, produced by the posterior part of the pituitary gland, that causes an increase in blood pressure but also decreases the flow of urine. Previously known as antidiuretic hormone.

vector   Any DNA that can propagate itself rapidly in a host and maintain this capability after insertion of foreign DNA into the vector. Although there are many types of vectors, the most common are derived from bacterial plasmids or the DNAs of bacterial and animal viruses. Most vectors in use today have been subjected to genetic engineering for specific purposes, such as the expression of foreign proteins in bacteria or animal cells (expression vectors).

vectorial discharge   Secretion of a substance by a cell at only one location or area on the cell surface. Examples of vectorial discharge include the secretion of mucin at the apical surface of epithelial cells lining the gut and the discharge of neurotransmitters into the synaptic cleft.

veratridine   A powerful neurotoxin derived from the lily, *Schoenocaulon officinalis,* used to study nerve function. Veratridine interferes with the action of sodium channels such that sodium ions can pass freely into the neuron.

very short patch repair (VSPR)   The type of excision repair that involves mismatches between single bases.

vimentin   A protein comprising one of the subclasses of intermediate filaments making up tissue-specific cytoskeletons in mammalian cells. Vimentin filaments are characteristically found in cells of mesenchymal origin (derived from embryo tissue), such as fibroblasts.

vinblastine   An antitumor drug isolated from the Madagascar periwinkle plant, *Vinca rosea.* Vinblastine acts to cause depolymerization of microtubules by binding to the tubulin protein subunits.

vincristine   An antitumor drug isolated from the Madagascar periwinkle plant, *Vinca rosea;* a chemical variant of vinblastine that acts to cause depolymerization in the same manner as VINBLASTINE.

vinculin   A cell surface protein localized to the adhesion plaque. Whereas the function of vinculin is unknown, alterations in the synthesis of the protein that accompany transformation suggest that this protein may play a role in the regulation of cell growth. (See EXTRACELLULAR MATRIX.)

viomycin   An antibiotic that acts by blocking a critical step in protein synthesis (translocation), thereby causing the polypeptide synthesis to be blocked before completion. Viomycin is itself a peptide that binds to the large or small ribosomal subunit to block translocation.

virion   A complete virus particle that includes the viral nucleic acid (either RNA or DNA) and, in some cases, enzyme molecules enclosed in a protein capsid.

viroid   An unusual infectious agent that produces diseases in plants and consists only of a naked, circular strand of RNA.

virulent, virulence   The property of a rapid spread of a pathogenic agent (e.g., a virus or bacteria) through a susceptible population.

virus   An agent that infects single cells but consists only of the components of the virion and does not possess the cellular machinery required for its own replication. For this reason viruses are, of necessity, intracellular parasites not clearly classifiable as living organisms.

virusoids   One of two classes of small infectious RNA molecules in plants. Virusoids do not contain genes for their own replication or packaging, but require a second, helper virus to accomplish these functions. Virusoids are also referred to as satellite RNAs.

viscosity   The property of resistance to flow exhibited by a substance in a fluid, semisolid state. In biochemical solutions viscosity is an indicator of a solution containing large macromolecules.

vitamin   An essential nutrient in the diets of mammals; an organic molecule that generally function as cofactors for specific enzyme-catalyzed reactions involved in energy production.

vitamin A   Any of the various chemical relatives of retinol (e.g., retinoic acid). Vitamin A is the precursor of the visual pigment, visual purple and has profound effects on the differentiation of epithelial tissues, including anticancer effects. Derivatives of vitamin A have been used as therapeutic agents for a variety of skin conditions, such as icthyosis, acne, and wrinkling.

vitamin B6   Any of various derivatives of pyridoxine (e.g., pyridoxal phosphate) used as a cofactor in the transamination reaction, the critical step by which amino acids enter the tricarboxylic acid cycle (Krebs cycle).

vitamin B12   A ring-shaped, cobalt-containing molecule also known as cobalamine. Vitamin B12 is an essential cofactor for the entry of certain amino acids and fatty acids into the tricarboxylic acid cycle (Krebs cycle).

W

Waldenstrøm's macroglobulinemia   A tumor of the lymphatic system characterized by oversecretion of IgM immunoglobins. Immunoglobulins derived from this tumor were used to derive the pentameric structure of IgM-type immunoglobulins.

Watson, James (b. 1928)   Along with Francis Crick, Watson demonstrated the double-helical structure of DNA by X-ray crystallography. This discovery showed that DNA had the requisite characteristics of a macromolecule that could serve as the genetic material.

Watson and Crick were awarded the Nobel Prize in medicine in 1962.

western blot A technique for identifying polypeptides that have been separated by polyacrylamide gel electrophoresis based on the reaction of specific antibodies to the proteins after they are transferred from the gel to an artificial membrane.

white matter That portion of the brain consisting of myelinated nerve fibers (axons) which carry nerve impulses from the gray matter of the brain and give it a characteristic white color.

wild-type gene The normal, nonmutated version of a gene.

wobble hypothesis The idea that, for an amino acid for which there is more than one triplet in the genetic code, the first base will always be the same in the different triplets but that the second and third positions of the triplets will vary by "wobble," with the third base of the triplet exhibiting the greatest wobble. (See CODON.)

writhing number (W) A number representing the turning of the axis of a supercoiled (DNA: $W = L - T$, where T = the twisting number.

X

xanthine-guanine phosphoribosyl transferase (gpt) gene (See HYPO-XANTHINE-GUANINE PHOSPHORIBOSYL TRANSFERASE.)

X chromosome One of the two sex chromosomes. The sex of the fetus is determined from the sex chromosomes present in the fertilized egg: two X chromosomes in the fertilized egg will produce a female, and one X chromosome and one Y chromosome will produce a male.

xenograft A graft from a foreign donor, such as human skin grafted onto a mouse.

xeroderma pigmentosum (XP) A family of genetic diseases in which excision repair mechanisms are faulty. This illness results in increased sensitivity to sunlight and, in particular, a much higher incidence of sunlight-induced cancers.

X-gal A synthetic substrate for the enzyme β-galactosidase (β-gal) that produces a blue product when acted upon by the enzyme. In vectors that carry the gene for β-gal, growth of bacteria containing a recombinant DNA, inserted into the β-gal gene (thereby inactivating the gene) by cloning, can be detected by the color of the colony they produce. (See INSERTIONAL INACTIVATION.)

X-linked diseases Genetic diseases carried on one of the sex chromosomes.

X-ray crystallography A technique for deducing the physical dimensions of a molecule (sizes of the atoms, lengths of the bonds between them) by examining how the path of an X-ray beam is altered as it passes through a crystallized sample of the molecule of interest. (See X-RAY DIFFRACTION.)

X-ray diffraction A technique for determining distances between atoms in a molecule by analyzing the diffraction pattern produced when an X-ray beam passes through molecules in a crystallized form.

Y

yeast(s) A subclass of fungi whose members are single celled. Yeasts display the major characteristics of higher cells, including chromosomes, an endoplasmic reticulum, and sexual mating. Because they are among the simplest eukaryotic cells with these characteristics, they are used as convenient and easily manipulable model systems to study the molecular genetics of eukaryotic cells.

yeast artificial chromosome A type of VECTOR used for cloning extremely large DNA fragments in yeast.

The vector is constructed by combining those elements of the yeast chromosome necessary for chromosome replication with the foreign DNA. The recombinant DNA created in this way can then be grown in a yeast host for many generations.

yes oncogene An oncogene found in Yamaguchi sarcoma virus and associated with sarcoma tumors in birds. The human proto-oncogene homologue has a protein kinase function that acts on the tyrosine amino acid residues.

Z

z-DNA A form of DNA in which the two strands are twisted around each other in a left-handed helix as opposed to the right-handed helix found in the more common form (i.e., B-DNA).

zinc finger A feature of many DNA-binding regulatory proteins in which a zinc atom is bonded to four amino acids (generally cysteine and histidine residues) so as to hold the polypeptide in a loop called a "finger." The finger is necessary for the DNA binding properties of the protein.

zoo blot A Southern blot in which a probe from a DNA suspected to represent a gene in one species is tested for its relatedness to sequences in other species. If the probe is found to hybridize to

DNAs from other species, then, because genes tend to be conserved in other species, this suggests that the probe represents a gene. (See SOUTHERN BLOT HYBRIDIZATION, HYBRIDIZATION STRINGENCY.)

zygote The developing embryo at any stage following the fusion of two gametes.

zygotic-effect genes Genes that effect the SEGMENTATION pattern of the embryo of the fruit fly, *Drosophila melanogaster,* that are derived from both the maternal and paternal parent as opposed to those that are active in the egg even before fertilization and so are referred to as maternal effect genes. (See SEGMENTS.)

APPENDIX I: ACRONYMS (AND OTHER ABBREVIATIONS)

A	adenine
ACP	acyl carrier protein
ACTH	adrenocorticotrophic hormone
ADP	adenosine diphosphate
AIDS	acquired immune dificiency syndrome
AMP	adenosine monophosphate
APC	adenomatous polyposis coli
APH	aminoglycoside-3'-phosphotransferase
araA	arabinosyladenine
araC	arabinosylcytosine
ARC	AIDS related complex
ARS	autonomously replicating sequences
ATP	adenosine triphosphate
AZT	3'-azido-3'-deoxythymidine
BAP	benzylaminopurine
BOD	biochemical oxygen demand
bp	base pair
5-BU	5-bromouracil
C	cytosine
cAMP	cyclic AMP
CAP	catabolite activator protein
CAT	chloramphenical acetyl transferase
ccc DNA	covalently closed circular DNA
Ccrit	critical dissolved oxygen concentration
cdc	cell division cycle
cDNA	complementary DNA
CF	complement-fixation
CFT	cystic fibrosis transmembrane conductance
CFTR	cysticic fibrosis transmembrane conductance regulator
CFU	colony-forming unit
cM	centimorgan
CML	chronic myelogenous leukemia
CMP	cytidine monophosphate
CNBr	cyanogen bromide
CoA/CoASH	coenzyme A
con A	concanavalin A
CRP	catabolite repression protein
CsCl	cesium chloride
ctDNA	chloroplast DNA
CTP	cytidine triphosphate
d	deoxy
DAG	diacylglycerol

dATP	deoxyadenosine triphosphate
dCTP	deoxycytidine triphosphate
dd	dideoxy
ddNTP	dideoxyribonucleotide triphosphate
dGTP	deoxyguanosine triphosphate
DHFR	dihydrofolate reductase
DMSO	dimethyl sulfoxide
DMT	dimethoxytrityl
DNA	deoxyribonucleic acid (See also ctDNA, mtDNA.)
DNase	deoxyribonuclease
dNTP	deoxyribonucleotide triphosphate
DP	docking protein
EBV	Epstein–Barr virus
ECM	extracellular matrix
E. coli	*Escherichia coli*
EDTA	ethylenediaminetetraacetate
EF	elongation factor
EGF	epidermal growth factor
ELC	expression-linked copy
ELISA	enzyme-linked immunoabsorbent assay
EMBL	European Molecular Biology Lab
EMS	ethylmethane sulfonate
ER	endoplasmic reticulum
erb	erythroblastosis
FACS	flourescence-activated cell sorter
f-actin	filamentous actin
FAD	flavin adenine dinuclestide
FBJ	Finkel, Biskis and Jinkins (discoverers of the FBJ murine osteosarcoma virus)
FCS	fetal calf serum
fes	feline sarcoma
FFU	focus-forming unit
FMN	flavin mononucleotide
FRA	*fos* related antigens
FRAP	fluorescence recovery after photobleaching
FSH	follicle stimulating hormone
FSV	feline sarcoma virus
ftz	fushi tarazu
5-FU	5-fluorouracil
G	guanine
G actin	globular actin
GAG	glycosaminoglycan
gal	galactosidase
GALT	gut associated lymphatic tissue
GC	gas chromatography
GDP	guanosine diphosphate
GFAP	glial fibrillary acidic protein
GH	growth hormone
GLC	gas-liquid chromatography
GMP	guanosine monophosphate

gpt	guanine phosphoribosyl transferase
GTP	guanosine triphosphate

HAT	hypoxanthine aminopterin-thymine
HCG	human chorionic gonadotropin
HDL	high-density lipoproteins
Hfr	high-frequency recombination strain
HGPRT	hypoxanthine-guanine phosphoribosyl transferase
HIV	human immunodeficiency virus
HLA	human leukocyte-associated antigens
HLTV	human T-cell leukemia virus
HMG	high-mobility group
HN	hemagglutinin-neuraminadase
hnRNA	heterogeneous nuclear RNA
HPFH	hereditary persistence of fetal hemoglobin
HPLC	high-performance liquid chromatography
HPV	human papilloma virus
HRP	horseradish peroxidase
HSE	heat shock response element
hsp	heat shock protein
HSR	homogeneously staining region
HSV	herpes simplex virus

IAA	indole acetic acid
IDL	intermediate-density lipoprotein
IF	initiation factor
IL	interleukin
IMP	inosine monophosphate
IPTG	isopropyl-β-D-thiogalactopyranoside
IR	infrared
IS	insertion sequence
IUdR	iododeoxyuridine

j gene	joining gene

k_D	diffusion coefficient/constant
Ki-MuSV	Kirsten sarcoma virus
k_M	Michaelis–Menten constant

LAV	lympho adenopathy virus
LDL	low-density lipoprotein
LH	lutinizing hormone
LINES	long-period interspersed sequences
LTR	long terminal repeat

MAPS	microtubule-associated proteins
MAR	matrix attachment regions
MAT	mating-type locus
MBP	maltose binding protein
MCP	methyl-accepting chemotaxis protein
mdr	multidrug resistance
5MeC	5-methylcytosine
MHC	major histocompatibility complex

MIF	migration-inhibitory factor
MMTV	mouse mammary tumor virus
MPF	M-phase promoting factor
mRNA	messenger RNA
mtDNA	mitochondrial DNA
MuLV	murine leukemia virus
myb	myeloblastosis
myc	myelocytomatosis

NAD^+	nicotinamide adenine dinucleotide
NADP	nicotinamide adenine dinucleotide phosphate
NANA	N-acetylneuraminic acid
NBT	nitro blue tetrazolium
N-CAM	neural cell adhesion molecule
NGF	nerve growth factor
NK cells	natural killer cells
NMR	nuclear magnetic resonance
NP40	nonidet P40
NTG	neomycin, thymidine kinase, glucocerebroside

oc	open circle

PAGE	polyacrylamide gel electrophoresis
papova virus	papilloma, polyoma, and vacuolating viruses
PAS	periodic acid-Schiff stain
PCR	polymerase chain reaction
PDGF	platelet-derived growth factor
PE	phosphatidylethanolamine
PEG	polyethylene glycol
PEP	phosphoenol pyruvate
PFU	plaque-forming unit
PITC	phenyl isothiocyanate
PKU	phenylketonuria
PML	progressive multifocal leukoencephalopathy
PMU	polymorphonuclear leukocyte
poly U	poly uridylic acid
PPLO	pleuropneumonia-like organisms
PVP	polyvinylpyrrolidone

raf	rat fibrosarcoma
ras	rat sarcoma
RBC	red blood cell
RER	rough endoplasmic reticulum
RES	reticuloendothelial system
RFLP	restriction fragment length polymorphism
RG	resorufin-β-D-galactopyranoside
RNA	ribonucleic acid (See also hnRNA, mRNA, rRNA, snRNA, tRNA.)
RNP	ribonucleoprotein
ros	Rochester 2 sarcoma
rRNA	ribosomal RNA
RSV	Rous sarcoma virus
RVE	reconstituted viral envelope

SAM	S-adenosylmethionine
SAR	scaffold attachment regions
SDGF	sarcoma-derived growth factor
SDS	sodium dodecyl sulfate
SEM	scanning electron microscopy
sis	simian sarcoma
snRNA	small nuclear RNA
SRP	signal-recognition particle
STS	sequence tagged site
SV40	simian virus 40
T	thymine, twisting number
Taq	*Thermus acquaticus*
TB	tuberculosis
Tc	cytoxic T cell
TCA	tricarboxylic acid
TEM	transmission electron microscope
TGF	transforming growth factor
Ti	tumor inducing
TMV	tobacco mosaic virus
TNF	tumor necrosis factor
tPA	tissue plasminogen activator
TPA	12-O-tetradecanoylphorbol-13-acetate
TRH	TSH-releasing hormone
tRNA	transfer RNA
TSH	thyroid stimulating hormone
Ts mutant	temperature-sensitive mutant
TTP	thymidine triphosphate
Ty	transposon yeast
U	uracil
UTP	uridine triphosphate
UV	ultraviolet
VLDL	very low density lipoprotein
VSG	variable-surface glycoprotein
VSPR	very short patch repair
W	writhing number
XP	xeroderma pigmentosum
YAC	yeast artificial chromosome

APPENDIX II: THE GENETIC CODE

UUU	} phenylalanine	UCU		
UUC		UCC	} serine	
UUA	} leucine	UCA		
UUG		UCG		
CUU		CCU		
CUC	} leucine	CCC	} proline	
CUA		CCA		
CUG		CCG		
AUU	} isoleucine	ACU		
AUC		ACC	} threonine	
AUA		ACA		
AUG	methionine	ACG		
GUU		GCU		
GUC	} valine	GCC	} alanine	
GUA		GCA		
GUG		GCG		

UAU	} tyrosine	UGU	} cysteine	
UAC		UGC		
UAA	} STOP	UGA	STOP	
UAG		UGG	tryptophan	
CAU	} histidine	CGU		
CAC		CGC	} arginine	
CAA	} glutamine	CGA		
CAG		CGG		
AAU	} asparagine	AGU	} serine	
AAC		AGC		
AAA	} lysine	AGA	} arginine	
AAG		AGG		
GAU	} aspartate	GGU		
GAC		GGC	} glycine	
GAA	} glutamate	GGA		
GAG		GGG		

Stop = translation stop signal

APPENDIX III: PURINE AND PYRIMIDINE BASES FOUND IN NUCLEIC ACIDS

Purines

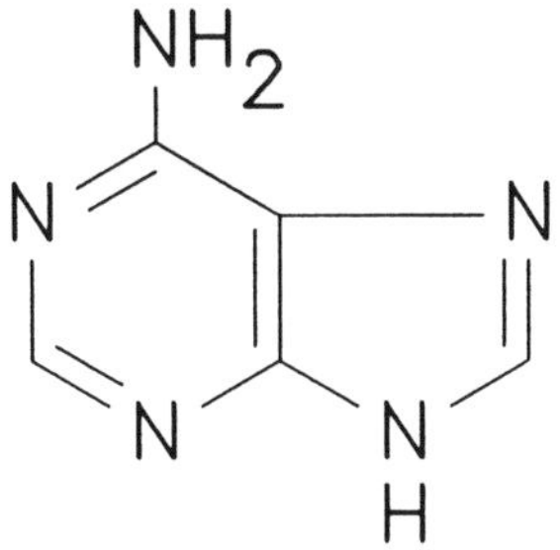

adenine

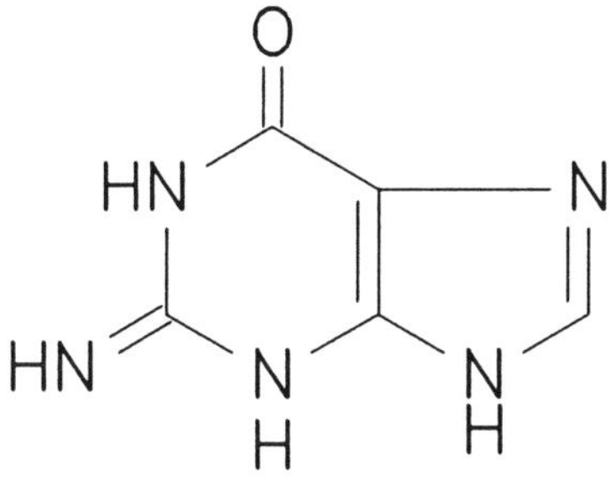

guanine

Pyrimidines

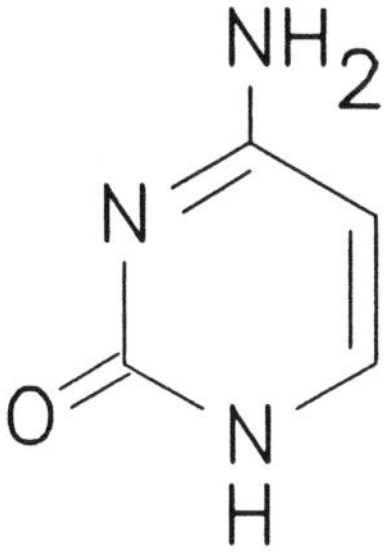

cytosine

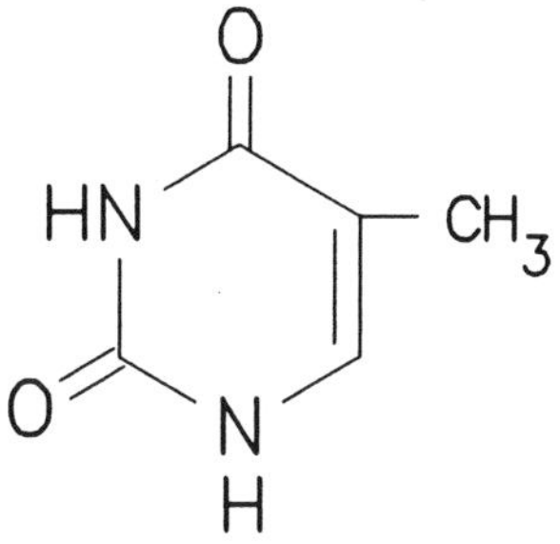

thymine (DNA only)

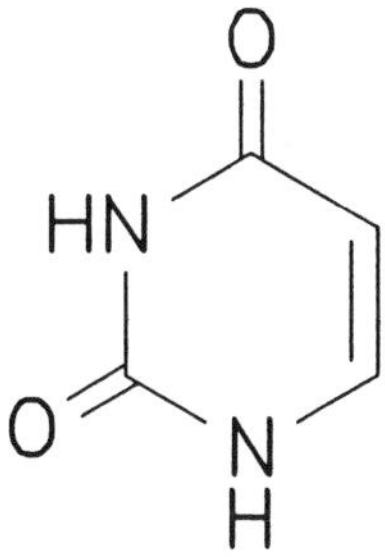

uracil (RNA only)

APPENDIX IV: SIDE CHAINS (R GROUPS) FOR INDIVIDUAL AMINO ACIDS

amino acid backbone

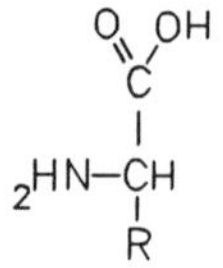

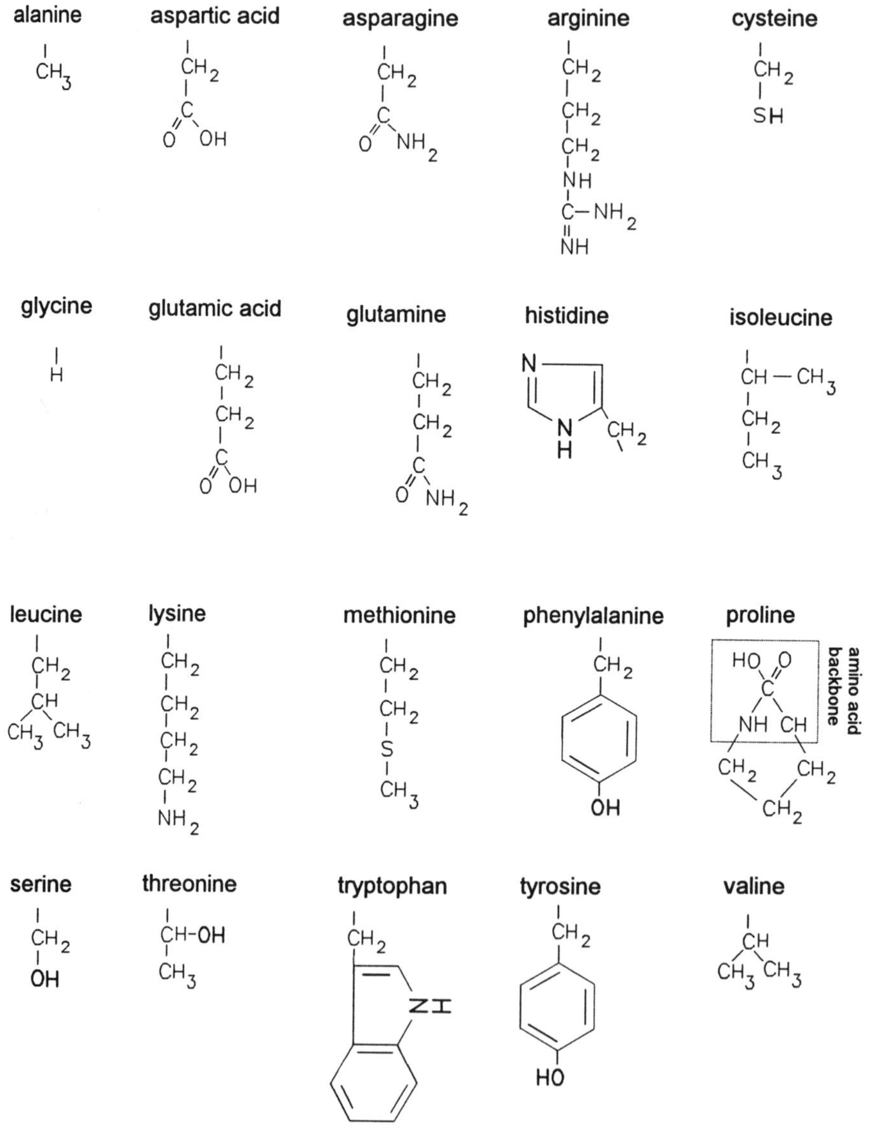

alanine	aspartic acid	asparagine	arginine	cysteine

glycine	glutamic acid	glutamine	histidine	isoleucine

leucine	lysine	methionine	phenylalanine	proline

amino acid backbone

serine	threonine	tryptophan	tyrosine	valine